PRINCIPLES OF FIRE BEHAVIOR

James G. Quintiere, Ph.D.

Australia • Brazil • Japan • Korea • Mexico • Singapore • Spain • United Kingdom • United States

Principles of Fire Behavior
James G. Quintiere

Publisher: Alar Elken

Acquistions Editor: Mark Huth

Developmental Editor: Jeanne Mesick

Production Coordinator: Toni Bolognino

Art and Design Coordinator: Michelle Canfield

Editorial Assistant: Dawn Daugherty

Marketing Coordinator: Mona Caron

Library of Congress Control Number: 97-11199

ISBN-13: 978-0-8273-7732-5

ISBN-10: 0-8273-7732-0

Delmar
Executive Woods
5 Maxwell Drive
Clifton Park, NY 12065
USA

Cengage Learning is a leading provider of customized learning solutions with office locations around the globe, including Singapore, the United Kingdom, Australia, Mexico, Brazil, and Japan. Locate your local office at **international.cengage.com/region**

Cengage Learning products are represented in Canada by Nelson Education, Ltd.

For your lifelong learning solutions, visit **www.cengage.com/delmar**

Visit our corporate website at **www.cengage.com**

Printed in the United States of America
14 15 16 17 18 11 10

To my parents who taught me never to play with fire and more; and to my sons, Chris and Scott, who kept me in play and more.

Contents

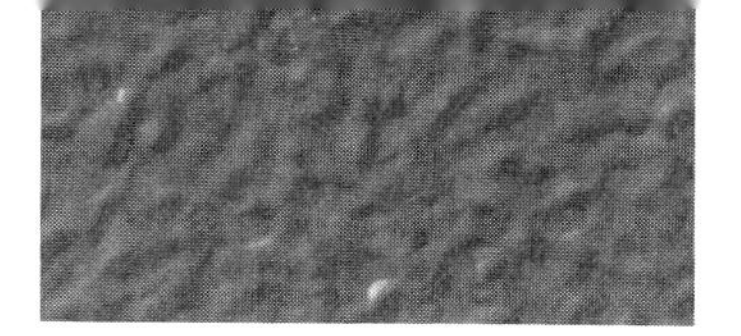

Preface

This book has been written for the practitioner in fire—the firefighter, code official, or investigator. A working knowledge of algebra is needed to effectively use the formulas presented. Acquired technical knowledge or course background in the sciences will be helpful. The text contains material given to fire protection engineers, but not at the same depth of theory or detail given in an engineering course. However, the conceptual explanations contained here may benefit these more advanced students as well.

The book arose from course material prepared in conjunction with training sessions for the Bureau of Alcohol, Tobacco, and Firearms (BATF) arson programs. Parts of it have also been used in a BATF course for state and local officials under the auspices of the Federal Law Enforcement Training Center (FLETC). I have benefited greatly from this association by enriching my scope of real fire events and in appreciating the application of science to them. Rick Miller and Jackie Herndon of BATF encouraged me to meet the educational challenge of these programs, and Bill Petraitis (BATF) has helped to convince me of the benefits of fire analysis to investigators.

This book begins with a perspective on the fire problem, the background of research, and how the student needs to approach the study of fire. Visualization is important to display concepts, and mathematical symbols and scientific units need to be digested.

An overview of fire addresses diffusion flames, premixed flames, smoldering, and spontaneous combustion. Experiments with a candle flame are described to help define some concepts.

Heat transfer, ignition, flame spread, and burning rate are discussed. Formulas are presented to allow the student to examine and appreciate quantitative aspects of fire. Energy release rate ($\dot{Q}$) in kilowatts (kW) is shown to be an important fire parameter.

Later chapters explain the fire smoke plume, products of combustion, and how fire behaves in a room. Again, quantitative formulas are presented.

The book closes with a discussion of the use of fire analysis to fire safety design and fire investigation. Examples are given, including a description of the Branch Davidian fire near Waco, Texas (April, 19, 1993). Hopefully by the final chapter the student will see how to apply the book material to his or her problems.

A cautionary note to the inexperienced reader. Fire can be unexpectedly injurious and deadly. A small flame can cause a burn injury in seconds, and a fire in a home can become deadly and inescapable in minutes. The experiments discussed and suggested for educational purposes, if conducted, need to be in an environment where attention has been given to proper ventilation and control of combustibles. Extreme care should be given in the use and handling of liquid fuels.

Acknowledgments

The author and Delmar Publishers would like to thank the following reviewers for the comments and suggestions they offered during the development of this project. Our gratitude is extended to:

Edward M. Garrison, CFI
ATF
Raleigh, NC

Robert Laford
Massachusetts Firefighting Academy
Orange Fire Department
Orange, MA

Michael McKenna
Mission College
Santa Clara, CA

Timothy G. Pitts
Guilford Technical Community College
Greensboro Fire Department
Greensboro, NC

Rod Westerfield
Training Horizons
Southern Stone County Fire Protection District
Kimberling City, MO

Theodore K. Cashel
Fire Marshal
Princeton Township
Princeton, NJ

Clint Smoke
Northern Virginia Community College
Annandale, VA
Chris Hawley
Baltimore County Fire Department
Towson, MD

David L. Fultz
LSU Fireman Training Program
Baton Rouge, LA

Tom Harmer
Titusville Fire & Emergency Services
Titusville, FL

Jack Fenner
University of Cincinnati
Cincinnati, OH

John Mirocha
Senior Special Agent
Bureau of Alcohol, Tobacco, and Firearms
Chicago, IL

The Evolution of Fire Science

Learning Objectives

Upon completion of this chapter, you should be able to:

- Define fire.
- Develop an awareness of fire in history.
- Understand the factors motivating fire research.
- Recognize the role of visualization and scale models in fire research.
- Explain S. I. units and scientific symbols.

INTRODUCTION

To understand fire, we must have a scientific definition of fire consistent with our perceptions. We must understand the role fire has played in history—its benefits and its costs to society in terms of people and property damage. Controlled fire, or combustion, for useful power is studied in conjunction with the market forces that drive our economy. The study of uncontrolled fire appears to be motivated by clear risks to society and by societies having the means to invest in such study. The development of the science of fire has accelerated over the last 150 years. It is a complex area involving many disciplines, and it is relatively primitive compared to other technological fields.

WHAT IS FIRE?

fire
an uncontrolled chemical reaction producing light and energy sufficient to damage the skin

combustion
fire, or controlled fire

Before there was life there was **fire**. It has left its imprint on history in many ways. In scientific terms, fire or **combustion** is a chemical reaction involving fuel and an oxidizer—typically the oxygen (O_2) in the air. Rusting and the yellowing of old newsprint do fit this definition; however, those processes are neither combustion nor fire for energy must be released for there to be either. Can we distinguish combustion from fire? In scientific terms, combustion and fire are synonymous. In conventional terms, we generally treat fire as distinct from combustion, in that fire is combustion that is not intended to be controlled. Firefighters attempt to control it by adding water or other agents, but the process of fire is not "designed" combustion, as in a furnace or an engine. Combustion experts who study such systems may know very little about fire, and those who deal with fire may know very little about combustion.

Fire is a chemical reaction that involves the evolution of light and energy in sufficient amounts to be perceptible. Will there always be light in a flame (fire)? No. For instance, the burning of hydrogen (H_2) with air or oxygen produces only water vapor from its chemical reaction. Although significant energy is produced, we would not see flame. But in most other fire classifications we would see the combustion process and a fire or a flame would be energetic enough to be sensed, particularly with sufficient energy to damage our skin. It may not be very big, but its energy release rate per unit volume of the chemical reaction zone would be sufficient to give us a local burn injury. This is fire. It is the result of striking a match, the glow of the charcoal briquette, the conflagration of the forest, and the spontaneity of a smoking haystack.

One last word on this chemical reaction called fire: On Earth it typically involves hydrocarbon-based fuels, those composed of atoms of carbon (C), hydrogen (H), and perhaps some oxygen (O) and nitrogen (N). Man-made substances have been added to this array of materials (molecules), such as chlorine (Cl), bromine (Br), fluorine (F) and other atoms. For example, wood molecules consist

■ NOTE
A chemical reaction destroys molecules but not atoms.

of the atoms involving H, C, and O; polyvinyl chloride (plastic) contains H, C, and O plus Cl atoms; and polyurethane (plastic) contains H, C, and O plus N atoms. These additions to the H-C-O base complicate the nature of combustion products and their potential threat to the environment.

All chemical reactions conserve mass, which means all the atoms survive, in contrast to a nuclear reaction in which some atoms are converted into new atoms with some matter transformed to energy. In a chemical reaction, however, molecules are not conserved. Their destruction is the essence of a chemical reaction in which they are converted to new molecules. For combustion or a fire, the formation of new molecules from the fuel and oxygen molecules gives off a net amount of energy. This energy comes from releasing the binding forces that hold the molecules together.

This is as close as we get to molecular physics. From here on, our discussion of fire principally concerns what we can see and feel. Of course we rely on some measurements, but these are at the macroscopic level in contrast to the microscopic or molecular level.

NATURAL CAUSES OF FIRE

Natural phenomena that can cause fire because of their high temperatures are lightning and molten rock from volcanic activity. These phenomena date back to the formation of the Earth. As organic matter developed, we could expect to have fire. Even today, these phenomena are a leading cause of accidental fire. Lightning strikes are recorded by the workers keeping watch over our forests. These strikes can cause smoldering underbrush on the forest floor, and a day or two after a storm, flames can erupt. The forest crews track these strikes and attempt to follow up with needed extinguishment. Of more concern to modern society is the effect of an earthquake, rather than direct volcanic activity. An earthquake can play havoc with fire and fuel sources used for heating and cooking. More damage was done by the conflagrations resulting from the San Francisco earthquake of 1906 and the Great Kanto earthquake of 1923 in Tokyo and Yokohama than from the actual quakes. During the fire following the Kanto earthquake, 38,000 people were killed by a fire whirl—a flaming tornado—that spun off the main fire column. These people had taken refuge in a park adjacent to the Sumida river in the Hitukuso-Ato district of Tokyo. This horror is depicted in the Japanese print in Figure 1-1a and in the gruesome photograph of corpses resulting from the fire whirl in Figure 1-1b. It is surprising that more study is not done on the effects of fire due to earthquake as opposed to the structural impact of the jolt.

There is one more natural cause of fire that humankind has not experienced, but unfortunately, the dinosaurs did. It has been generally established that in about 65 million B.C. a large meteorite smashed into the Earth. The meteorite's pas-

Figure 1-1a *Japanese painting of the fire whirl during the great Kanto earthquake in Tokyo, 1923. Photo courtesy of Y. Hasemi.*

Figure 1-1b *Photograph of corpses from the effect of the fire whirl in the Hitukuso-Ato district of Tokyo, 1923. Photo courtesy of K. Saito.*

sage through the atmosphere caused frictional heating, and its impact (equaling many nuclear bombs) propelled debris and fire products into the atmosphere resulting in a soot cloud that circled the Earth and affected the atmosphere for at least a year, causing a perpetual winter and the demise of the dinosaurs by virtually eliminating their food supply. This cataclysmic event probably occurred more than once, sealing the fate of the dinosaurs. The consequences of this fire event are known because remnants of the meteorite (iridium) have been found in various parts of the world in the strata of rock from the same geologic time: between the Cretaceous and Tertiary periods.

The so-called nuclear winter has the potential to destroy life on Earth, just as a perpetual winter destroyed the dinosaurs. This winter effect of a large fire was publicized in the 1980s when nuclear warfare studies determined that the fire from a limited nuclear war could produce debris and smoke that would result in a nuclear winter. Our world would be in jeopardy from a sustained reduction of light and temperature. Many studies were done to substantiate this "fallout" from even a survivable nuclear war. The noted science writer, Carl Sagan, was one of the strongest advocates for studying the consequences of nuclear winter. Even before the cold war ended, all nuclear parties were chilled by this possibility.

■ **NOTE**

Aristotle considered fire sufficiently important to classify it as one of the four elements of matter: Fire, earth, air, and water.

Long after the dinosaurs, humankind evolved and eventually cultivated fire. Aristotle thought it sufficiently important to classify it as one of the four elements of matter: Fire, earth, air, and water. Fire was used and abused in many ways. Hazel Rossotti vividly describes the uses, hazards, and spiritual qualities of fire as used by society.[1] She describes how Cherokee Indians learned to preserve fire by burying a smoldering log, then digging it up and fanning it into flames. This cyclical process probably came as a surprise to those who experienced the Stump Dump Fire near Baltimore, Maryland, in 1990. This incident involved more than 5 acres of buried tree stumps. After many other attempts at extinguishment failed, the stumps were more completely buried, probably slowing the smoldering, but not extinguishing it.

FIRE IN THE UNITED STATES

History is punctuated with fire disasters. The Great Fire of London (1666) and the Chicago Fire (1871) caused the destruction of thousands of buildings. These were literally forest fire-like events in that wind was a principal factor in their spread. Once the wind died, the fire stopped. Mechanical suppression apparatus could not deal with such fires until calm conditions prevailed. These fires were **conflagrations** or **mass fires** involving large tracts at a given time. Their lateral flame extent is much greater than their flame height. Such fires were induced in World War II in the bombing raids of Hamburg, Dresden, and Tokyo, which were more destructive than the effects of the atomic bombs on Hiroshima and Nakasaki. Indeed, many of the greatest disasters in recent history from natural phenomena are likely to be due to fire (see Figure 1-2).

conflagration or mass fire
a fire over a large tract of land where the flames are generally much shorter than the horizontal extent of the fire

Type and Location	Number of Deaths	Date of Disaster
Floods:		
Galveston, Tx. tidal wave	6,000	Sept. 8, 1900
Johnstown, Pa.	2,209	May 31, 1889
Ohio and Indiana	732	Mar. 28, 1913
St. Francis, Calif., dam burst	450	Mar. 13, 1928
Ohio and Mississippi River valleys	380	Jan. 22, 1937
Hurricanes:		
Florida	1,833	Sept. 16–17, 1928
New England	657	Sept. 21, 1938
Louisiana	500	Sept. 29, 1915
Florida	409	Sept. 1–2, 1935
Louisiana and Texas	395	June 27–28, 1957
Tornadoes:		
Illinois	606	Mar. 18, 1925
Mississippi, Alabama, Georgia	402	Apr. 2–7, 1936
Southern and Midwestern states	307	Apr. 3, 1974
Ind., Ohio, Mich., Ill., and Wis.	272	Apr. 11, 1965
Ark., Tenn., Mo., Miss., and Ala.	229	Mar. 21–22, 1952
Earthquakes:		
San Francisco earthquake and fire	452	Apr. 18, 1906
Alaskan earthquake-tsunami hit Hawaii, Calif.	173	Apr. 1, 1946
Long Beach, Calif., earthquake	120	Mar. 10, 1933
Alaskan earthquake and tsunami	117	Mar. 27, 1964
San Fernando—Los Angeles, Calif., earthquake	64	Feb. 9, 1971
Marine:		
"Sultana" exploded—Mississippi River	1,547	Apr. 27, 1865
"General Slocum" burned—East River	1,030	June 15, 1904
"Empress of Ireland" ship collision—St. Lawrence River	1,024	May 29, 1914
"Eastland" capsized—Chicago River	812	July 24, 1915
"Morro-Castle" burned—off New Jersey coast	135	Sept. 8, 1934
Aircraft:		
Crash of scheduled plane near O'Hare Airport, Chicago	273	May 25, 1979
Aircraft (Continued):		
Crash of scheduled plane, Detroit, Mich.	156	Aug. 16, 1987
Crash of scheduled plane in Kenner, La.	154	July 9, 1982
Two-plane collision over San Diego, Calif.	144	Sept. 25, 1978
Crash of scheduled plane, Ft. Worth/Dallas Airport	135	Aug. 2, 1985
Railroad:		
Two-train collision near Nashville, Tenn.	101	July 9, 1918
Two-train collision, Eden, Colo.	96	Aug. 7, 1904
Avalanche hit two trains near Wellington, Wash.	96	Mar. 1, 1910
Bridge collapse under train, Ashtabula, Ohio	92	Dec. 29, 1876
Rapid transit train derailment, Brooklyn, N.Y.	92	Nov. 1, 1918
Fires:		
Peshtigo, Wis. and surrounding area, forest fire	1,152	Oct. 9, 1871
Iroquois Theatre, Chicago	603	Dec. 30, 1903
Northeastern Minnesota forest fire	559	Oct. 12, 1918
Cocoanut Grove nightclub, Boston	492	Nov. 28, 1942
North German Lloyd Steamships, Hoboken, N.J.	326	June 30, 1900
Explosions:		
Texas City, Texas, ship explosion	552	Apr. 16, 1947
Port Chicago, Calif., ship explosion	322	July 18, 1944
New London, Texas, school explosion	294	Mar. 18, 1937
Oakdale, Pa., munitions plant explosion	158	May 18, 1918
Eddystone, Pa., munitions plant explosion	133	Apr. 10, 1917
Mines:		
Monongha, W. Va., coal mine explosion	361	Dec. 6, 1907
Dawson, N. Mex., coal mine fire	263	Oct. 22, 1913
Cherry, Ill., coal mine fire	259	Nov. 13, 1909
Jacobs Creek, Pa., coal mine explosion	239	Dec. 19, 1907
Scofield, Utah, coal mine explosion	200	May 1, 1900

Source: World Almanac, National Transportation Safety Board, National Weather Service, National Fire Protection Association, Chicago Historical Society, American Red Cross, U.S. Bureau of Mines, National Oceanic and Atmospheric Administration, and city and state Boards of Health.

Figure 1-2 *Life loss due to national disasters in the United States. Courtesy National Safety Council,* Accident Facts, *1995 Edition (Ref. 2).*

Such enormous disasters are relatively infrequent. It is the local effects of fire that have contributed to our perception of its hazard. Approximately 2.5 million fires are reported in the United States each year. This suggests a fire each year for at least 1 in 50 households. Roughly 5,000 deaths occur each year in the United States due to fire. For a total population of 250 million with a life expectancy of 70 years, we calculate

$$\frac{250 \times 10^6 \text{ people}}{70 \text{ years} \times 5000 \text{ deaths per year}} = 714 \text{ people}$$

NOTE
Roughly 1 in 700 people will die by fire over your lifetime.

or roughly 1 in 700 people will die by fire in the United States over your lifetime. This fatality frequency is roughly 10 times more, or 1 in 70, than for automotive deaths. You may have a neighborhood fire about once a year, and you may know at least one person during your lifetime who will die by fire. These fire statistics relative to other threats do not cry out for a great concern about fire.

However, fires and statistics do move people to action. Throughout the twentieth century much effort has been put into the prevention of and protection from fire. In the United States, following the Great Fire of Baltimore (1904) and the fact that the property loss due to fire was then ten times that of Europe, a national research program was initiated at the National Bureau of Standards (NBS) to investigate the consequences of fire on building construction materials. Steel and concrete buildings were not necessarily found to be "fireproof." The NBS fire program in 1913 was established under Simon H. Ingberg who would head the Fire Resistance Section of the Heat Division for the next 40 years. In 1914 due to a Congressional mandate, funding, and the challenge of the complex fire problem, "so broad became the scope of the investigation that it soon involved almost every one of the scientific and engineering laboratories of the Bureau."[3]

In addition, the National Fire Protection Association (NFPA) was established in 1896 to provide a voluntary basis for technical information and standards—recommended procedures and practices—concerning fire safety. The Underwriters' Laboratory (UL) and the American Society of Testing and Materials (ASTM) contributed standard test methods to assess performance in fire conditions. This led to fire resistance tests, that enable the measurement of the endurance of structural elements of buildings and structures in fire. Such tests, conducted in standard furnaces throughout the industrialized world, help to ensure that structures do not collapse because of fire. But all fires do not match the conditions of the standard fire resistance furnace tests, and practices are not universal in their attempts to match the testing results to actual anticipated fire conditions. More striking is the fact that "standard" furnaces do not always impart the same fire heating conditions. Standard practices can have some imperfections.

NOTE
In the early 1970s, it was acknowledged that the United States had the highest annual death rate due to fire of the world's industrialized nations.

In the early 1970s, it was acknowledged that the United States had the highest annual death rate from fire of the world's industrialized nations. Current statistics for industrialized countries are listed in Table 1-1.

Table 1-1 *Annual fire death rates.*

Country	Annual Deaths per 10^5 persons	Country	Annual Deaths per 10^5 persons
Russia*	10.60	France	1.26
Hungary	3.31	Czech Republic	1.21
India*	2.20	Germany	1.17
Finland	2.18	Australia	0.93
Union of South Africa*	2.00	New Zealand	0.92
United States	1.95	Spain	0.86
Denmark	1.64	Poland*	0.80
Norway	1.60	Austria	0.74
Canada	1.58	Netherlands	0.63
Japan	1.52	Switzerland	0.53
United Kingdom	1.49	Italy*	0.30
Belgium	1.47	China*	0.20
Sweden	1.35		

Sources: From Wilmot, Ref. 4, for 1989–1992. Starred items are from Brushlinsky et al., Ref. 5 for 1994.

flashover
a sudden event in a room fire, leading to full involvement

The U.S. death rate over the last century is shown in Figure 1-3 with trends in other accidental death categories. The United States still ranks high, but the death rate has steadily decreased over the century. This fact could be attributed to better fire safety, but note that the total accidental death has decreased also.

Congress passed the Fire Prevention and Control Act of 1974. This act structured the U.S. Fire Administration to enhance the practice of firefighting, to improve education, to assess national fire statistics, and to develop research. It focused research on the development of fire in buildings: ignition, flame spread, smoke, and **flashover** (a sudden event in fire growth that rapidly leads to full involvement of a room). The safety of people rather than structures became the focus. This attitude of mandated public safety, coupled with the threat of civil suits because of fire, has motivated the increasing use of innovative fire safety technologies, including residential smoke detectors, smoke control systems in large buildings, and the extension of sprinklers to public and residential occupancies.

Changing technologies bring improvements in fire safety but can also bring new risks. The high death rate due to fire in the United States and in other industrialized nations may be associated with the changing technologies with which we live. Change brings risks, some unexpected. For example, in the 1970s, lightweight

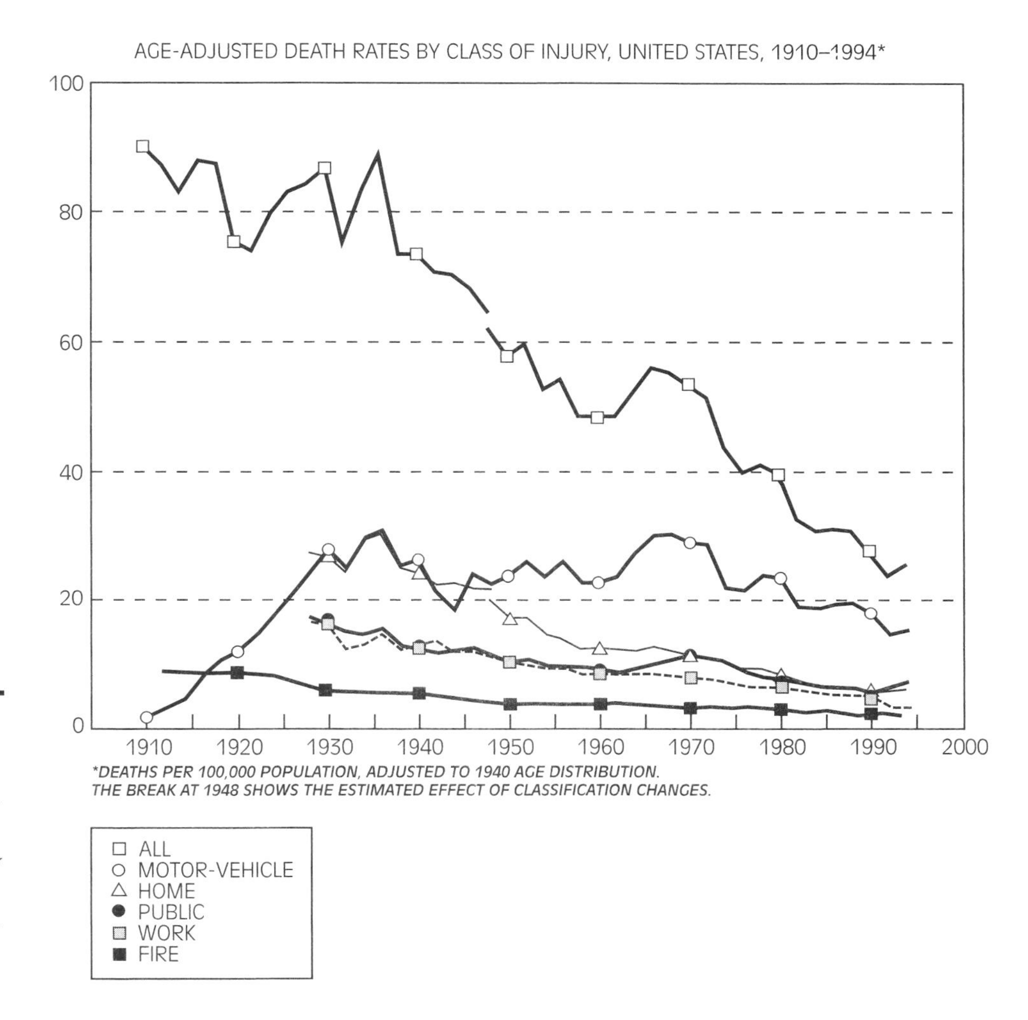

Figure 1-3 *The U.S. death rate over the last century, along with trends in other accidental death categories. Courtesy National Safety Council,* Accidental Facts, *1995 edition (Ref. 2).*

NOTE
The high death rate due to fire in the United States and in other industrialized nations may be associated with the changing technologies with which we live.

cellular plastics (such as foam polyurethane and polystyrene) were seeing new applications, but their fire hazards were not fully appreciated. Standard tests suggested no problems, but accident scenarios showed a dramatically fast rate of fire spread, which brought standard flammability tests into question. The Product Research Committee (PRC), established in 1974 was created to examine this issue for cellular plastics as a result of a consent order between the Federal Trade Commission and industry.[6]

Although the PRC group gained much insight into the issue of foam plastic flammability, it remains unresolved today. To mitigate concern, today lightweight cellular plastics are normally required to be covered by wallboard in building construction applications even after they pass the test requirements for flammability. Confidence in their flammability ratings by standard tests is low. But this issue of product flammability is not unique to cellular plastics, it is widespread. The disparity among flammability tests for construction materials is illustrated in Figure 1-4 for six national European tests in the ranking of twenty-four materials. This chart was publicized by Professor Howard Emmons after he discovered it in

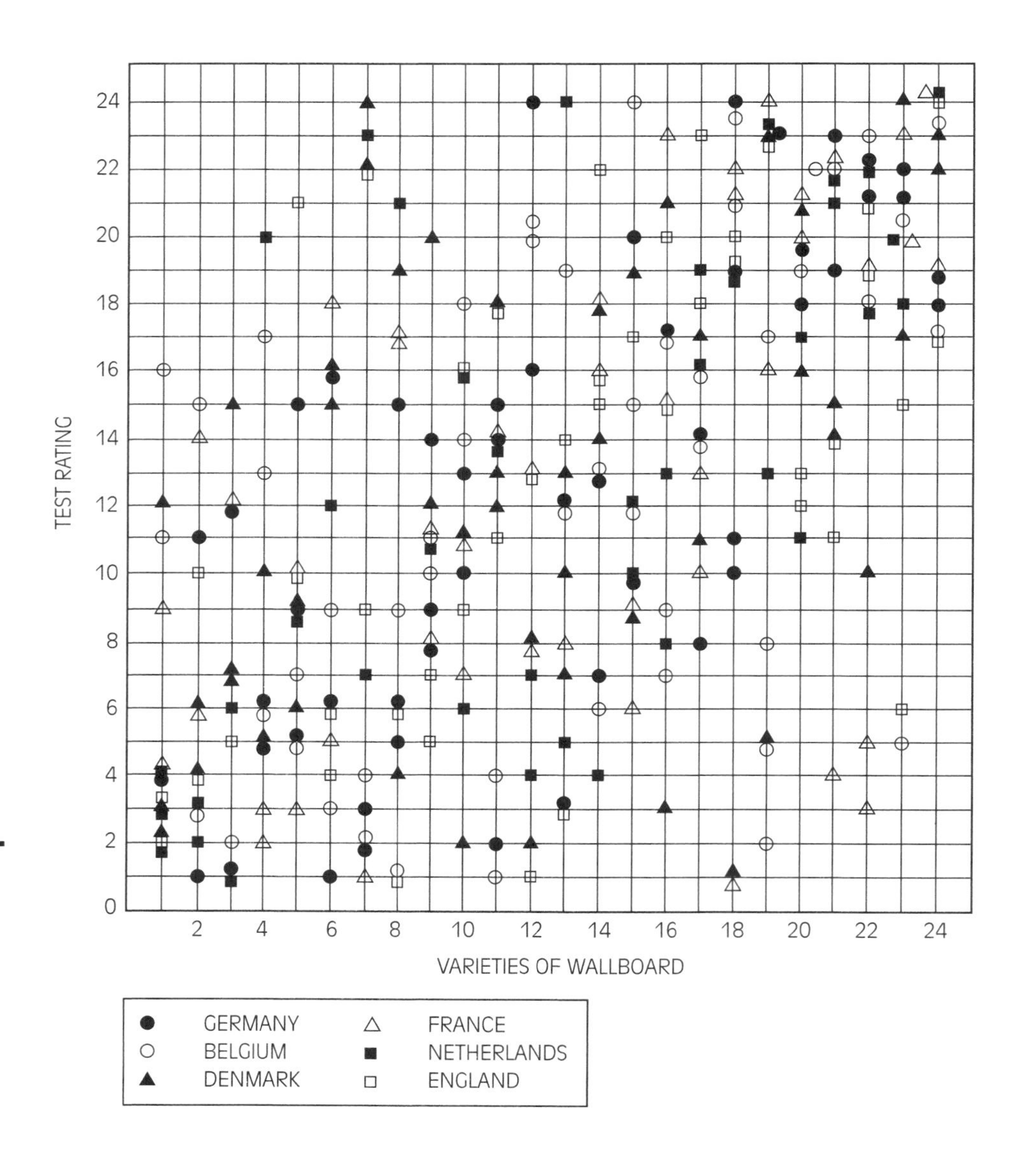

Figure 1-4
Disparities among fire tests—rankings of twenty-four materials by six national flammability test methods. From Emmons, Ref. 7.

Europe during a world tour to review the status of fire science.[7] A perfect correlation among all of the six tests should produce the same ranking for the twenty-four materials, a 45° straight line correlation. For example, the number 6 material should be ranked as 6 by all the tests for a perfect correlation of flammability. The tests decidedly do not correlate! The results are not consistent, giving ambiguity to the term *flammability*. Perhaps it is appropriate that in the book intended to present accurate prose (*The Elements of Style* by W. Strunk Jr. and E. B. White), the word *flammable* is listed among those misused:

> **Flammable.** An oddity, chiefly useful in saving lives. The common word meaning "combustible" is *inflammable*. But some people are thrown off by the *in-* and think *inflammable means* "not combustible." For this reason, trucks carrying gasoline or explosives are now marked FLAMMABLE. Unless you are operating such a truck and hence are concerned with the safety of children and illiterates, use inflammable.[8]

We do not currently have a universal test procedure to establish flammability or, alternatively, inflammability. These tests are reflections of our misuse of the word.

■ NOTE
It has been estimated that the annual cost of fire in the United States is $85 billion.

The annual costs of fire are not insignificant, but because of the relatively low frequency of fire compared to other societal threats, they are not fully recognized. In fact, statistics are likely to be inaccurate and incomplete. Moreover, fire safety costs do not increase the productivity of the economy; they are a drag on the economy. When they become too great, more attention will be made to minimizing them. It has been estimated that the annual cost of fire in the United States is $85 billion.[9] This total includes property loss, business interruption, product liability, insurance administrative costs, fire service (paid), and fire protection in construction and equipment. Table 1-2 shows the percentage of major costs—property loss ($10 billion, U.S.A.), fire protection in construction ($20 billion, U.S.A.), fire service (~$10 billion, U.S.A.) and insurance administrative costs (~$6 billion, U.S.A.)—for a range of countries scaled to their gross domestic products (GDP). As can be seen, it is typical for many developed countries to invest nearly 1% of its GDP in fire-related costs. It could be as high as 2% if more complete accounting were done. Unfortunately, not even 1% or 2% of these costs are invested in improving our knowledge of the technology of fire safety. Lessons are not just learned from history, but must also be analyzed using the science of fire.

■ NOTE
It is typical for a country to invest nearly 1% of its GDP for fire-related costs.

■ NOTE
Future disasters will shape the course of fire safety.

Nevertheless, future disasters will shape the course of fire safety. In developing nations, high-rise buildings are being built to heights that dwarf current levels. It is possible that thousands of occupants will be seriously affected or killed by a fire disaster in a high-rise building. Future living arrangements in outer space under low gravity and in underground and undersea structures will present unexpected fire hazards. Fire affecting radioactive operations and toxic waste storage sites will present hazards of new dimensions. Despite improvements in fire safety technologies, there will be new surprises in the technological advancements of society.

Table 1-2 *National annual fire costs by percentage of gross domestic product for 1989 to 1991.*

Country	Property Loss	Building Fire Protection	Fire Insurance Admin.	Fire Fighting	Total
Hungary	0.12	0.42	0.01	ND	ND
Spain	0.12	ND	0.05	ND	ND
Japan	0.14	0.29	0.11	0.27	0.81
Finland	0.17	ND	0.05	0.18	ND
U.S.A.	0.17	0.37	0.07	0.28	0.89
Canada	0.18	0.18	0.21	0.16	0.73
New Zealand	0.18	0.14	0.23	0.20	0.75
West Germany	0.19	ND	0.09	ND	ND
Netherlands	0.20	0.12	0.04	0.17	0.53
Austria	0.21	ND	0.14	ND	ND
United Kingdom	0.22	0.19	0.10	0.27	0.78
Switzerland	0.23	0.29	ND	ND	ND
Denmark	0.28	ND	0.08	0.09	ND
Sweden	0.28	0.10	0.06	0.19	0.63
France	0.29	0.14	0.16	ND	ND
Norway	0.31	0.26	0.15	0.12	0.84
Belgium	0.40	0.21	0.28	0.18	1.07

Source: From Wilmot Ref. 4.
ND = data not available.

FIRE RESEARCH

The study of fire is a complex subject that comprises an array of interdependent disciplines. Each of these subjects needs to be developed before the pieces can be put together to adequately describe fire. Science is the evolution of many steps and contributions. Eventually the subject takes shape, and individuals formulate or unify the subject by quantitative description that allows for predictions and assessments into a formally recognized scientific discipline. Let us briefly examine the history of these component subjects.

thermodynamics
the study of energy and states of matter

Fire is uncontrolled combustion involving chemistry, thermodynamics, fluid mechanics, and heat transfer. **Thermodynamics**, the study of energy and states of matter, was principally shaped by Willard Gibbs, a noted nineteenth century sci-

heat transfer
the transport of energy from a high- to low-temperature object

fluid mechanics
the study of fluid motion

entist, who brought a unification and clarity to the subject that is still appreciated today. **Heat transfer** also had its roots in the early 1800s. Joseph Fourier, a general in Napolean's army, formulated the law of heat conduction that forms the theoretical basis of the field. But heat transfer in fluids had to await the development of modern **fluid mechanics** in the late 1800s when O. Reynold's pioneering work on turbulent flow laid the basis for engineering analysis. In the 1900s, Theodore von Kármán and others advanced the subject of aerodynamics, which paved the way for a more complete framework for heat transfer. At this point, solutions were based on approximate methods, since the governing mathematical equations were too complex to exactly solve. In combustion, Y. B. Zel´dovitch, a Russian scientist, was able to formulate solutions for diffusion flames by innovative approximate mathematical techniques. Although begun in the 1930s, the subject of combustion was not developed to a mature state until the 1950s. Today, large computers make it possible to examine many facets of the subject but issues of turbulence, chemical kinetics, and other small-scale phenomena still cannot be completely resolved by computer solutions. The engineer must still rely on intellectual insight and approximate formulations. Moreover, only when a thorough understanding of a subject is mastered can simple representations of complex phenomena be made. This also allows the transfer of knowledge and its ease of use.

The subject of fire needed to build on all of its component disciplines. These had to mature before it was even possible to adequately describe and predict fire.

Another factor influencing the development of fire science is the motivation to study it in the first place. We have seen that fire is a drain on the economy, and there is no direct market incentive for its study.

In Japan, the consequences of earthquakes led to an extreme sensitivity to fire safety and its study. Fire science is studied in schools of architecture in Japan as well as in other scientific fields. It is endemic in Japanese culture and academic disciplines. Not surprisingly, the first science-based handbook on the quantitative description of fire was published in Japan in the early 1980s.

England developed one of the most advanced scientific laboratories for fire study in the world (the Fire Research Station, formerly at Borehamwood). Much of the work in the 1960s under the leadership of Dennis Lawson, Philip Thomas, and David Rasbash has not been fully appreciated because it was never published in mainstream fire journals. But it is recognized that the first graduate program in fire engineering was founded by David Rasbash at the University of Edinburgh, and Philip Thomas has been a main force in disseminating the benefits of fire science throughout the world.

Many of the pioneers of fire science came together in a special meeting in Washington, D.C. on November 9 and 10, 1959, for "The Use of Models in Fire Research." Walter Berl, the conference organizer, commented, "The intimate interplay between aerodynamics, heat transfer, and chemical reaction rates makes the study of fires the intriguing problem it is."[10] Scientists who have achieved prominence outside the field of fire have elected to study fire because of its challenge and the prospect of its benefits to society.

In the United States, one of the earliest scientists of fire was Hoyt Hottel of M.I.T. He began the study of fire before World War II, but was diverted to study the effects of fire from weapons during the war. Later, he and Emmons of Harvard University pursued fundamental research in fire and lobbied for government research funding. Such research support was realized in the early 1970s by expansions of fire research at Factory Mutual Research Corporation and at the National Institute of Standards and Technology (formerly the National Bureau of Standards). It was made possible by targeted funding for fire research by the National Science Foundation and through the National Fire Prevention and Control Act of 1974. The fruits of that research effort in the 1970s helped to promote, develop, and catalyze the disconnected efforts of fire research throughout the world. Although U.S. funding for basic research in fire has since decreased, the worldwide activity is expanding in its communication links, and there is a healthy exchange of knowledge in this small field. For example, proceedings of the symposia sponsored by the International Association for Fire Safety Science help to maintain international exchange in fire research. It is this synthesis of fire research that makes possible this book. The *SFPE Handbook of Fire Protection Engineering*[11] developed by the Society of Fire Protection Engineers (SFPE) is a good illustration of the current knowledge base of fire science compiled by experts among the disciplines of fire.

VISUALIZATION OF FIRE PHENOMENA

■ **NOTE**
The shape of a flame is influenced by the fluid flow induced by the flame itself.

It is crucial to have a visual concept of fire phenomena before one can establish a framework for study. Many effects are seen (or can be seen if planned) during the progression of fire and its related smoke movement. These must be categorized if we are to learn. The shape of a flame is influenced by the fluid flow induced by the flame itself. The nature of smoke movement in buildings can take many forms. Such visualization must take place in the laboratory for systematic study, but firefighters and others must be able to articulate their observations to scientists to promote their study. The scientist cannot allow the size of his laboratory to limit the scope or relevance of his observations in fire phenomena. However, small-scale studies can be very relevant. We all appreciate the role of wind tunnels in the design of aircraft and in improving the aerodynamics of motor vehicles. The Wright brothers had a wind tunnel at their disposal. Such scale modeling techniques pervade many fields of study; fire is not excluded. Physical scale models based on the laws of science can help to design aircraft, ships, oceanic tidal basins, concert halls, and even analyze vehicle crash dynamics. Processes in fire can also be reasonably represented by scale models. Many of the formulas we will study in this book have resulted from scaling techniques using laboratory-size systems.

Figure 1-5 is a schematic illustration of phenomena arising in a room fire. The recognition of these and other phenomena helped to establish a framework for the "modeling" of compartment fires. Figures 1-6 and 1-7, respectively, show the dynamics of smoke movement and fluid flow in a corridor subjected to a room

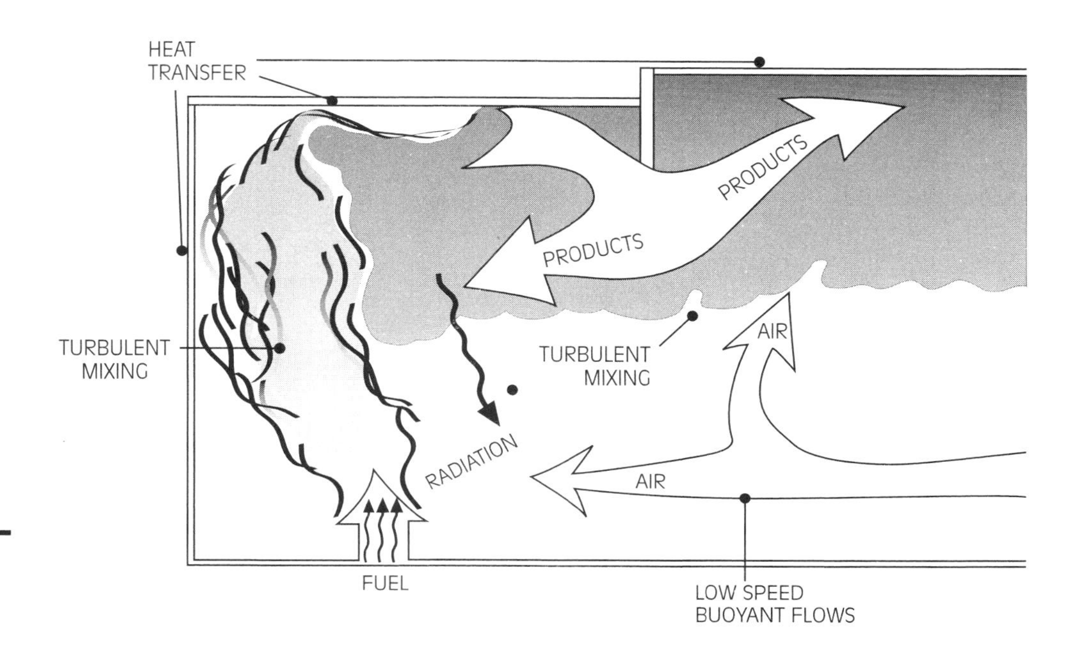

Figure 1-5
Phenomena in a room fire growth.

fire.[12] The results were accurately established using a scale model with glass walls and colored smoke to make the processes visible. These photographs show the classical representation of an apparent uniform smoke layer filling the upper half of a compartment (Figure 1-6) and the complex recirculating flows that occur both in the smoke layer and the relatively clear space below (Figure 1-7). Figure 1-6 also illustrates the mixing of smoke into the lower clear space (at the right) as air enters through a window and entrains smoke.

Figure 1-8 shows fascinating flame patterns as a result of flames impinging on a ceiling.[13] The patterns depend on fuel type, fuel flow rate, and spacing to the ceiling. No form of computer modeling can anticipate these results.

SCIENTIFIC NOTATION

In order to relate concepts and principles behind fire phenomena, we must first qualitatively understand the behavior of fire. More importantly, we must understand the information and methods used to quantitatively describe specific fire phenomena. But some knowledge of physics and algebra is needed. Some basics of units of measure, symbols, and scientific notation will provide the foundation for our quantitative analyses.

International System of Units or Standard International (SI) Units the system of units for measurement adopted by the science community

In today's world, despite the lagging of the United States, the **International System of Units (SI)** is in widespread use. The scientific community has univer-

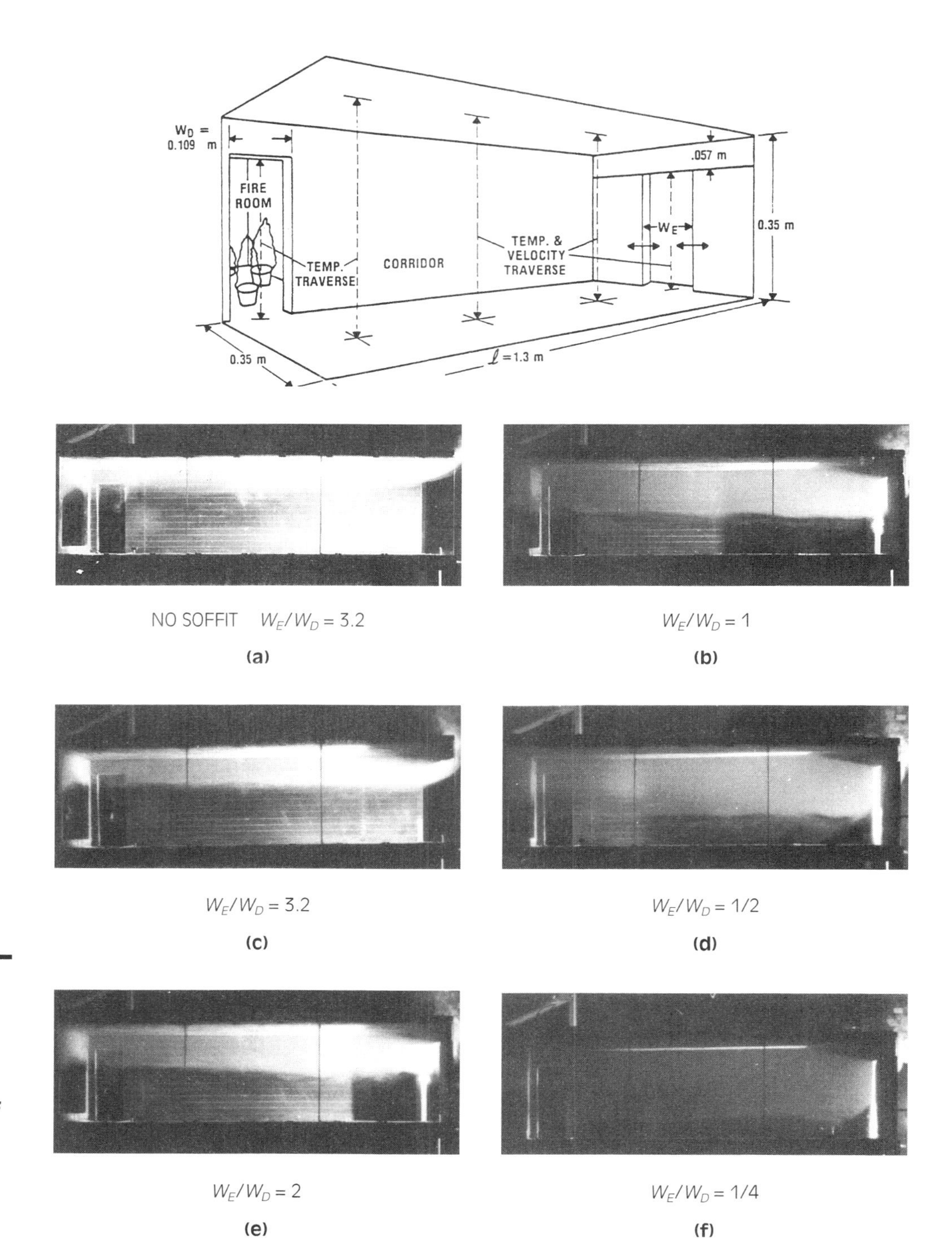

Figure 1-6 *Smoke layer in a scale model corridor subject to a room fire (W_D = 11 cm) at the left and exits through doorway of width (W_E) at the right. From Quintiere et al., Ref. 12.*

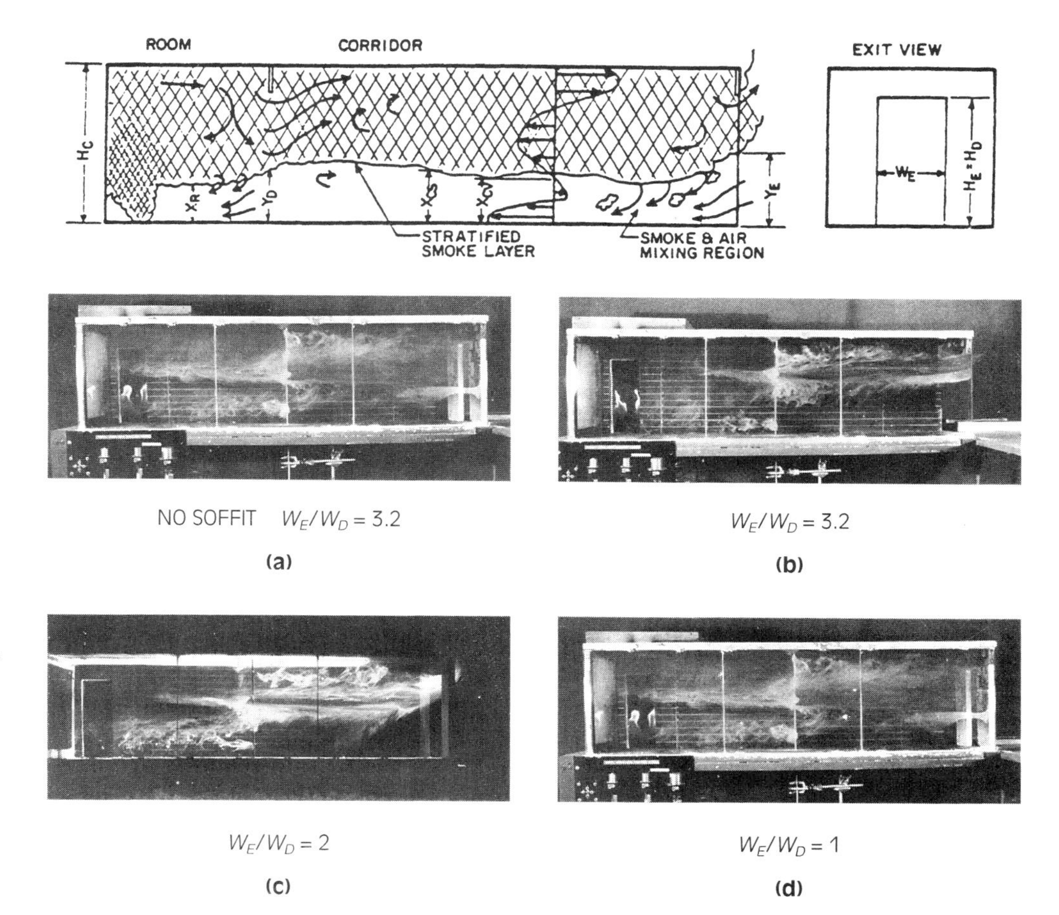

Figure 1-7 *Smoke streaks showing the complex flow pattern in a scale model corridor subject to a room fire. From Quintiere et al., Ref. 12.*

sally adopted this system in its publications. To understand articles in fire science, it is essential to be conversant with the SI units. Table 1-3 lists the SI units for quantities relevant to our forthcoming discussion. The quantities listed arise from the component disciplines of fire and cannot be developed in great depth, but we will come to them.

As an example, let us consider energy. Table 1-4 is related to temperature, and in the United States we should be familiar with BTU units (British Thermal Units) and perhaps W-s (Watt-second) which is a Joule (J).

Although temperature is more commonly expressed in Fahrenheit (°F), we can readily convert to other units of temperature as shown in Table 1-5. Celsius or Centigrade (°C) is based on water freezing and boiling at 0 and 100 respectively, whereas 32 and 212 are respectively assigned on the Fahrenheit scale. Both the °C and °F scales do not start their zero base where all thermal energy stops (absolute

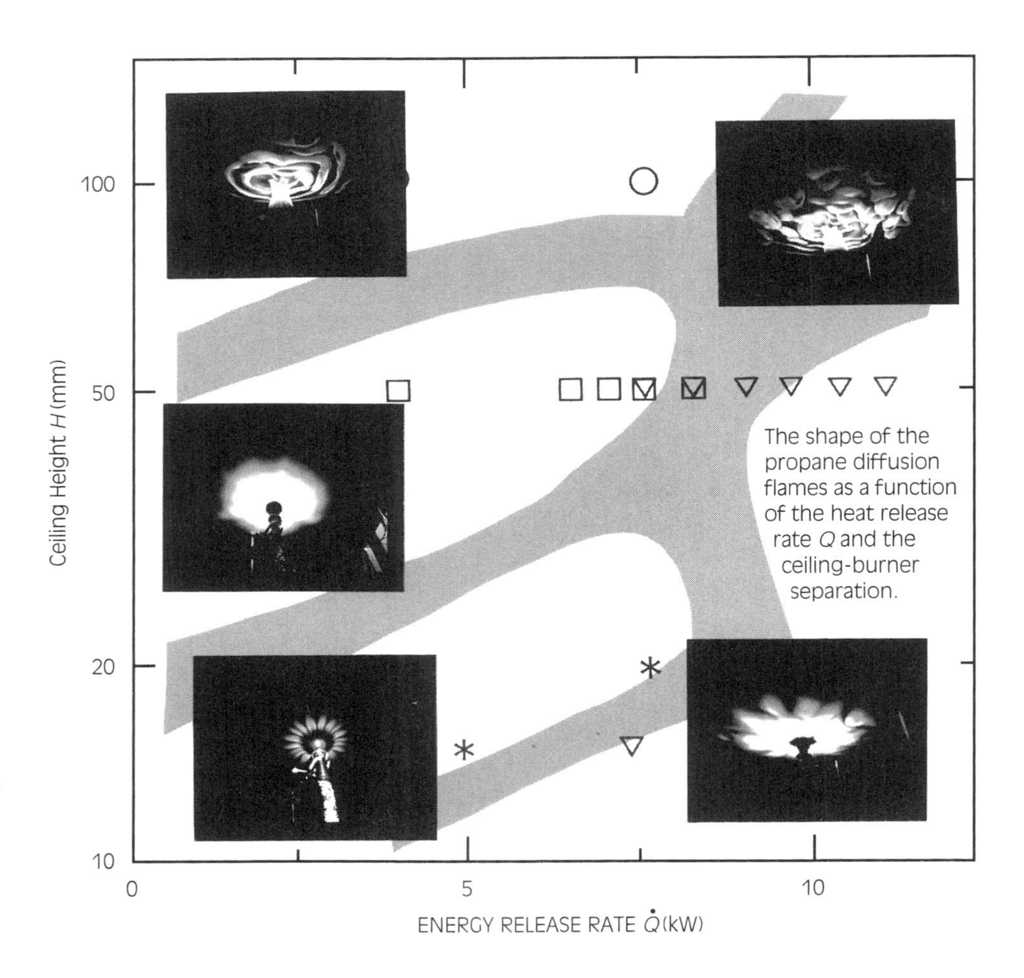

Figure 1-8 *Flame patterns of a ceiling jet. From Kokkala and Rinkinen, Ref. 13.*

zero). This is presented in alternative (absolute) scales of Rankine for °F and Kelvin for °C. Temperatures need to be in "absolute" units for certain formulas.

Table 1-6 gives other conversion factors that may be useful. In addition, typically used symbols are assigned to the quantities. These symbols are used in this text and are fairly common in the fire literature, but are not universal. Note that the dot over a symbol implies "rate" or "per unit time", and two accents implies "per unit area." Rate per unit area is commonly called **flux**. Greek symbols are also common, e.g. ρ (rho) for density and α (alpha) for thermal diffusively. We will explain the significance of these qualities as we encounter them in our study of fire.

flux
pertains to mass or heat flow rates per unit area

Finally Table 1-7 lists terminology in scientific notation which avoids many zeros in expressing numbers. For example, kW (kilowatts) denotes 1000 watts or 10^3 watts. Similarly 10^{-3} W is a thousandth of one watt or one mW (milliwatt).

Table 1-3 *SI quantities.*

Quantity	Unit abbreviation
Force	N (newton)
Mass	kg (kilogram mass)
Time	s (second)
Length	m (meter)
Temperature	°C or K
Energy	J (joule)
Power	W (watt)
Thermal conductivity	W/m - °C
Heat-transfer coefficient	W/m^2 - °C
Specific heat	J/kg - °C
Heat flux	W/m^2

Table 1-4 *Alternative energy units.*

1 Btu will raise 1 lb_m of water 1°F at 68°F.
1 cal will raise 1 g of water 1°C at 20°C.
1 kcal will raise 1 kg of water 1°C at 20°C.

Some conversion factors for the various units of work and energy are
1 Btu = 778.16 lb_f-ft
1 Btu = 1055 J
1 kcal = 4182 J
1 lb_f-ft = 1.356 J
1 Btu = 252 cal

Table 1-5 *Temperature conversions.*

°F degree Fahrenheit: T(F) = T(C) (1.8) + 32
°R degree Rankine: T(R) = T(F) + 459.69
°C degree Celsius or Centigrade: T(C) = (T(F)–32)/1.8
°K degree Kelvin: T(K) = T(C) + 273.16

Table 1-6 *Common conversion factors and symbols.*

length	1 m = 3.2808 ft	l
area	1 m^2 = 10.7639 ft^2	A
density	1 kg/m^3 = 0.06243 lb/ft^3	ρ
energy	1 kJ = 0.94783 Btu	Q
heat	1 kJ = 0.94783 Btu	q
heat flow rate	1 W = 3.4121 Btu/hr	$\dot{q}$
energy release rate	1 W = 3.4121 Btu/hr	$\dot{Q}$
heat flow rate per unit area, heat flux	1 W/cm^2 = 0.317 Btu/hr-ft^2 1 W/cm^2 = 10.kW/m^2	$\dot{q}''$
specific heat	1 kJ/kg-°C = 0.23884 Btu.lb-°F	c
thermal conductivity	1 W/m-°C = 0.5778 Btu/hr-ft-°F	k
thermal diffusivity	1 m^2/s = 10.7639 ft^2/s	α
pressure	1 atm = 14.69595 lb_f/in^2 = 1.01325×10^5 N/m^2 1 N/m^2 = 1 Pascal (Pa)	P

Table 1-7 *Scientific notations, prefixes.*

Multiplier	Prefix	Abbreviation
10^{12}	tera	T
10^{9}	giga	G
10^{6}	mega	M
10^{3}	kilo	k
10^{2}	hecto	h
10^{-2}	centi	c
10^{-3}	milli	m
10^{-6}	micro	μ
10^{-9}	nano	n
10^{-12}	pico	p
10^{-18}	atto	a

Note: For example, 10^3 = 1000 and 10^{-3} = 0.001.

Summary

Fire is a chemical reaction, usually involving oxygen from the air, that produces enough energy to be perceived. Typically, that energy release rate per unit volume is sufficient to cause a skin burn in seconds. The destruction by fire throughout history has been dramatic, from the possible cause of the dinosaur extinction to conflagrations in large cities into the twentieth century. Although, the United States has one of the highest per capita deaths due to fire, it is only approximately one-tenth of the annual death rate due to motor vehicle accidents. Consequently, there has not been a high incentive to study uncontrolled fire. However, research in fire has progressed throughout the world, and its results can be very useful to fire safety design and to fire investigation analysis. Visualization of fire phenomena is a first step in understanding and in developing tools for fire protection. To be able to use these tools, the student must understand scientific units and terminology. A goal in understanding this book is to at least come to think of fires in terms of kilowatts.

Activities

1. Find examples of the significance of fire in history.
2. Examine the statistics of fire and the attitude of society on fire safety.
3. Test the validity of fire statistics. For example, does the prevalence of electrical fire causes at night suggest something else? What percentage of fires are actually reported?
4. Examine the *SFPE Handbook*, the proceedings of the IAFSS (Fire Safety Science series) on topics of interest to you. Discuss how you can benefit from this science. Is it understandable?
5. Identify fire organizations and laboratories around the world.

Review Question

1. Convert the following:

85°F	_____ °C	30,000 BTU/hr	_____ kW
400°F	_____ °C	0.0015 W	_____ mW
695°F	_____ °C	1,385 W	_____ kW
		6.5 W/cm^2	_____ kW/m^2

References

1. H. Rossotti, *Fire* (Oxford: Oxford University Press, 1993), 22.
2. *Accident Facts* (Itasca, IL: National Safety Council, 1995).
3. R. C. Cochrane, *Measures for Progress, A History of the National Bureau of Standards* (Washington, DC: U.S. Department of Commerce, 1974), 131.
4. T. Wilmot, *World Fire Statistics Centre Bulletin 12* (Geneva: The Geneva Association, 1996).
5. N. N. Brushlinsky, A. P. Naumenko, and S. V. Sokolov, "Response to Query about Fire Deaths in Russia" (Letters to the Editor), *Fire Technology* 31, no. 3, (August, 1995).
6. *Fire Research on Cellular Plastics: The Final Report of the Products Research Committee*, Library of Congress Cat. No.80-83306 (1980).
7. H. W. Emmons, *Fire Research Abstracts and Reviews* 10, no. 2 (1968): 133.
8. W. Strunk, Jr., and E. B. White, *The Elements of Style* (New York: MacMillan, 1979), 47.
9. W. P. Meade, *A First Pass at Computing the Costs of Fire Safety in a Modern Society*, NIST-GCR-91-592 (Gaithersburg, MD: National Institute of Standards and Technology, March 1991).
10. W. G. Berl, ed., *The Use of Models in Fire Research*, Pub. 786 (Washington, DC: National Academy of Sciences, National Research Council, 1961), v.
11. P. J. DiNenno, ed., *The SFPE Handbook of Fire Protection Engineering*, 2d ed., (Quincy, MA: National Fire Protection Association, June 1995).
12. J. G. Quintiere, B. J. McCaffrey, and W. Rinkinen, "Visualization of Room Fire Induced Smoke Movement and Flow in a Corridor," *Fire and Materials* 2, no. 1 (1978): 18–24.
13. M. Kokkala, and W. J. Rinkinen, *Some Observations on the Shape of Impinging Diffusion Flames*, Res. Rep. 461 (Espoo, Finland: VTT, Technical Research Centre of Finland, 1987).

Additional Reading

Lyons, J. W., *Fire*, Scientific American Library (New York: Scientific American Books, 1985).

Fire Safety Science, Proceedings of the First–Fourth International Symposia. Symposia one and two, New York: Hemisphere Publishing Co.; Symposia three, London: Elsevier; Symposium four, International Association for Fire Safety Science.

Combustion in Natural Fires

Learning Objectives

Upon completion of this chapter, you should be able to:

- Identify different forms of natural fire: diffusion flames, spontaneous ignition, smoldering, and premixed flames.
- Understand how a candle flame, a basic diffusion flame, works.
- Understand, from a quantitative viewpoint, the natural forms of fire including aspects of size, shape, and speed.

INTRODUCTION

Natural fire processes can take different forms: diffusion flames, smoldering, spontaneous combustion, and premixed flames. By examining the candle flame with simple experiments derived from Michael Faraday's 1850 work, you can deduce how a diffusion flame works. It is also shown that the process of flaming ignition involves a premixed flame and a pilot flame. Spontaneous combustion requires no such pilot, but occurs because of itself and its environment. Spontaneous combustion can take either of two pathways: (1) flaming combustion—a diffusion flame or (2) smoldering combustion—a slow solid fuel oxidation at temperatures as low as 400°C. We describe these processes and present quantitative information.

FIRE AND ITS INGREDIENTS

Combustion or fire is a chemical reaction involving the release of energy, some of which is in the form of light—a flame. Most fuels are composed of carbon, hydrogen, and oxygen. Some fuels, particularly plastics, can contain other elements such as nitrogen, chlorine, and fluorine. To define a chemical reaction as fire, sufficient perceptible energy must be released: The rate of energy release per unit volume of the chemical reaction determines whether that reaction is fire. The size of the flame is not a factor. On the threshold of fire this incipient energy level might be about 10^3 or 1000 kW/m^3, which is sufficient to heat water 1°C per second. Sustained fire reactions can possess many more times this energy density—as much as 10^{10} kW/m^3. The temperature in this reaction zone can reach 2,000°C for gaseous fuels and 1,000°C for solid fuel reactions (smoldering).

fire triangle
a concept describing fire as consisting of three ingredients: fuel, oxygen, and energy

The **fire triangle**, as shown in Figure 2-1, is a concept used to describe the fire processes. The elements of the fire triangle are essential to the existence of a fire. The triangle consists of (1) fuel combining with (2) oxygen in a chemical reaction to release (3) energy and other chemical products. The energy causes heat to be transferred to the solid or liquid fuel to maintain vaporization into gaseous fuel or to maintain the fuel temperature to ensure the chemical reaction can persist. If we take away sufficient fuel or oxygen, or reduce the energy by extinguishment or retardant agents, the fire will not survive.

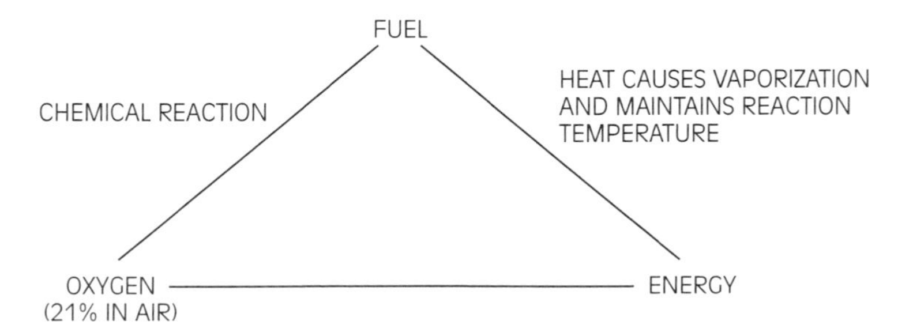

Figure 2-1 *The fire triangle.*

Types of Fire

We can categorize fire into four distinct phenomena:

1. Diffusion flames,
2. Smoldering,
3. Spontaneous combustion,
4. Premixed flames.

Diffusion flames represent the predominant category. It is the building fire, the forest fire, the lit match. Smoldering can be the birth or follow the death of a diffusion flame. It is the glowing embers of the tragic house fire, the result of a lightning strike to the forest bed, or the glow of a blown out match flame. Spontaneous combustion is the incubation of a chemical reaction that leads to smoldering or (diffusion) flaming. It can occur in oily cotton rags (e.g., linseed oil), a haystack, or a pile of wood chips. Premixed flames represent controlled combustion processes such as in the gasoline internal combustion engine with spark ignition or the diesel engine with autoignition. It also represents the incipient flame in the ignition of solids and liquids before a diffusion flame emerges.

diffusion flame
a flame in which the fuel and oxygen are transported (diffused) from opposite sides of the reaction zone (flame)

DIFFUSION FLAMES

A **diffusion flame** is a combustion process in which the fuel gas and oxygen are transported into the reaction zone due to concentration differences. This transport process is called **diffusion** and is governed by Fick's Law, which says that a given **species** (e.g., in connection with fire, oxygen, fuel, CO_2) will move from a high to low concentration in the mixture. A drop of blue ink in a glass of water will eventually diffuse into the water to give a blue tinge. Oxygen in air will move to the flame where it has a concentration of zero as it is consumed in the reaction. Fuel is transported into the opposite side of the flame by the same process. The combustion products diffuse away from the flame in both directions. This process is illustrated in Figure 2-2. Most natural flaming fires are diffusion flames. A common example is the flame of a match or a candle (Figure 2-3). In a candle the flame melts the wax, which is transported up the wick by capillary action. The flame then vaporizes the wax, and the gaseous fuel diffuses into the flame where it

diffusion
process of species transport from a high to low concentration

species
another name for distinct chemical compounds and other molecular structures in a mixture, usually gases

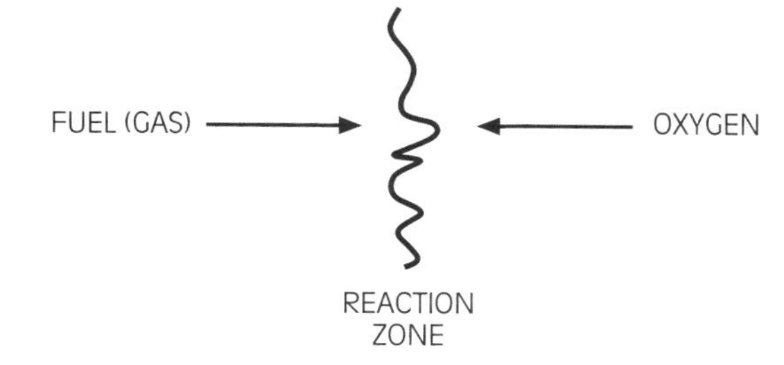

Figure 2-2
Schematic of a diffusion flame.

pyrolysis
a process of breaking up a substance into other molecules as a result of heating; also known as thermal decomposition

laminar
refers to orderly, unfluctuating fluid motion

turbulent
refers to randomly fluctuating fluid motion around a mean flow

buoyancy
an effective force on fluid due to density or temperature differences in a gravitational field

meets oxygen. For the wooden match the wood is decomposed by the heat of the flame into gaseous fuel and char. This decomposition process is called **pyrolysis**. A candle flame is an example of a **laminar** diffusion flame governed by pure molecular diffusion. Any flame higher than approximately 1 ft will naturally possess random fluid mechanical unsteadiness, illustrated by visible eddies in the smoke and flame. This is called turbulence, and so we have **turbulent** diffusion flames (Figure 2-4). Smoke from a large chimney has this turbulent character, and a smoke filament from a cigarette in a still room begins as a laminar flow then clearly breaks up after rising about 1 ft to become turbulent (Figure 2-5).

Gravity influences the shape of diffusion flames and profoundly affects fire processes in general. Because fire creates high temperatures, the hotter (lighter) gases rise as a result of **buoyancy** caused by gravity. The ensuing flow distorts the flame and eventually flow instabilities cause turbulence. Turbulence is due to disturbances, arising naturally, that excite the flow at its natural frequency. This same phenomenon occurs when an unbalanced tire begins to vibrate erratically at a speed associated with the wheel's natural frequency. The mechanical constraints restrict the wheel from ripping off the vehicle. Similarly, fluid friction keeps the erratic fluid's turbulent motion within limits. Accordingly, buoyancy and turbulence are two factors that control fire and its associated flows. It is interesting to contemplate the behavior of diffusion flames in a space ship in which gravity and therefore buoyancy is negligible.

The general shapes of diffusion flames are illustrated in Figure 2-6 on page 28. Natural fires involving liquid or solid fuels have very low velocities at the fuel

■ **NOTE**
Buoyancy and turbulence are two factors that control fire and its associated flow.

Figure 2-3 *A candle, an example of a diffusion flame.*

Figure 2-4
Turbulent diffusion flame.

jet flame
flame due to a high velocity fuel supply

base (~1 cm/s), but a high pressure gaseous fuel source can initiate fuel at much higher velocities. For such flames, buoyancy effects can be negligible, and at a high enough velocity, the **jet flame** reaches a fixed height. Fires with characteristics as shown in Figure 2-6b, more representative of normal burning commodities, achieve heights associated with their fuel supply and air supply drawn in by the buoyancy of the fire. For fires over large areas, such as forest fires or city conflagrations, the air flow can be drawn down from above as well as in from the sides. Figure 2-6b will be the focus of our attention and represents diffusion flames associated with building fires.

Figure 2-5 *Cigarette smoke.*

(c)

Figure 2-6 *Diffusion flame shapes: a. jet flame, b. liquid spill fire, c. forest fire.*

The Scientific Discoveries of Michael Faraday

1820 discovered two unknown chlorides of carbon

1821 discovered electromagnetic rotation—set up an experiment in which a wire carrying an electric current rotated in the field of a horseshoe magnet

1823 liquified chlorine

1825 isolated benzene

1831 demonstrated the principle of electromagnetic induction—the production of electric current by a change in magnetic intensity

1844 discovered the rotation of the plane of polarization of light in a magnetic field.

Figure 2-7 *Michael Faraday and his scientific discoveries. After Ref. 1.*

Candle Flame

The candle flame provides most of the essential features of natural fires and a diffusion flame. It will be our learning tool. We are not original in this approach because we shall repeat some of the experiments created by the nineteenth century scientist, Michael Faraday (Figure 2-7). These experiments were presented at the Royal Institution in London as a science show for the public. Known as the Christmas Lectures, they were very popular with children and have been archived in a book that is still available today.[1] Some of these lectures are remarkable in their simplicity, yet powerful in their illustration of the basic principles of flames.

It is astonishing to those who study fire to see how Faraday appreciated the workings of a candle and its implication to the science of fire. These thoughts are expressed in his opening remarks at the convening of the Christmas Lectures:

> I propose to bring before you, in the course of these lectures, the Chemical History of a Candle. There is not a law under which any part of this universe is governed which does not come into play and is touched upon in these phenomena. There is no better, there is no more open door by which you can enter into the study of natural philosophy than by considering the physical phenomena of a candle.[1]

Light a candle and observe the processes that create the sustained flame. As illustrated in the sketch in Figure 2-8, you can see several things. The yellow and blue zones constitute the flame (the former, diffusion; the latter, premixed). The wick is actually designed to curve so that the flame "clips" off the wick, and limits its height. Why isn't the wick destroyed throughout the flame? What is the purpose of the wick? With some thought, it can be deduced that the heat of the flame melts the wax, the melt soaks the wick (by capillary action, like water feeding tree leaves), evaporates, and supplies the fuel gas to diffuse into the luminous zone where it finds oxygen having diffused from the other side. Buoyant flow elongates

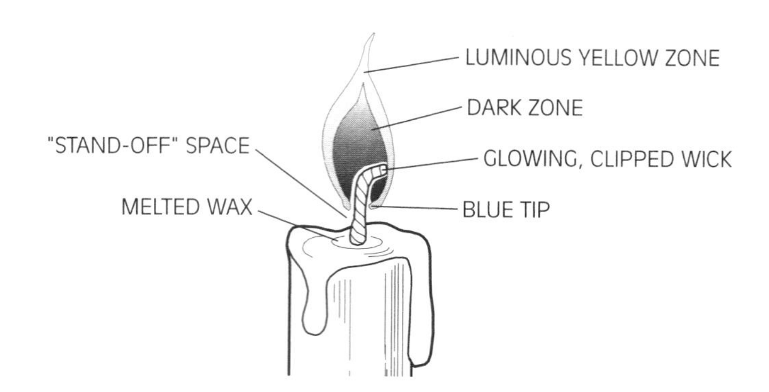

Figure 2-8 *Characteristics of a candle flame.*

the candle flame, but the flow path is small enough so that laminar flow is maintained. However, as the hot combustion gases rise from the flame, they become turbulent. This phenomenon can be seen by projecting a collimated light (from a slide projector or strong flashlight) at the flame and onto a white screen. You will see a shadow image of the flow on the screen as a result of the refraction of light rays caused by the hot and cold flows. Note that the filament-like plume from the flame becomes turbulent at about 1 ft above the flame. If you observe more closely, you will see the shadow of the wick and the shadow of a part of the flame region. What might cause this flame shadow? (Some solid ingredients blocking the light in the flame.)

■ **NOTE**
Combustion reactions can only be sustained if the flame temperature is high enough, typically greater than 1300°C.

Now place a metal screen (typical window screen mesh, but not aluminum) into the flame as shown in Figure 2-9. The metal screen cuts the flame, extinguishing further combustion. Why? Describe the luminous flame cut.

Combustion reactions can only be sustained if the flame temperature is high enough, typically greater than 1,300°C. As the height of the screen is varied in the

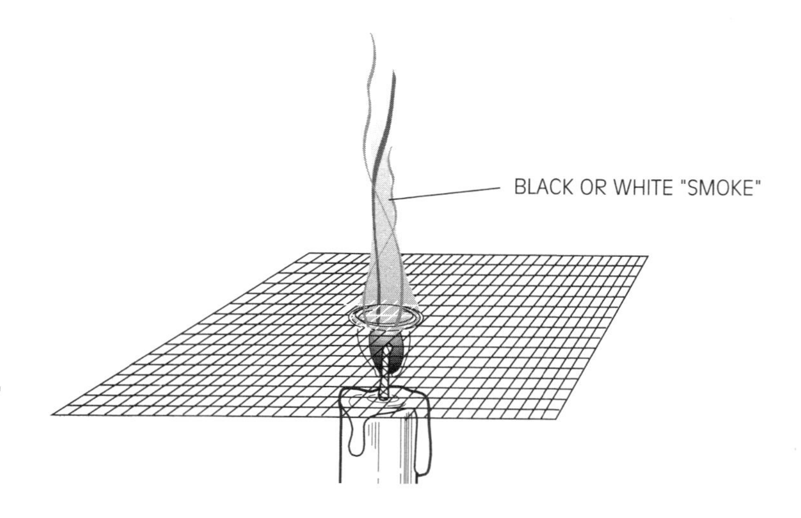

Figure 2-9 *Metal screen cutting flame.*

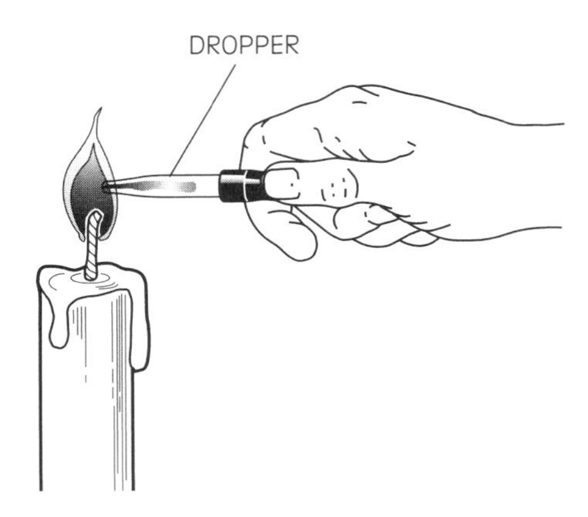

Figure 2-10
Extracting interior flame gases.

flame, black smoke is released with the screen high in the flame and white or grey "smoke" with the screen near the top of the wick. What do these colored smokes imply? Collect the black smoke on a piece of white paper, being careful not to insert the paper in the flame. What is collected on the paper? Where do these black particles originate? Find that region in Figure 2-8. Blow out the flame and observe the color and character of the smoke that momentarily emanates from the wick. How does this compare to the white smoke resulting from the low screen position?

Relight the candle. After the flame steadies, insert a glass dropper into the region where you think the white smoke originated. Extract the white smoke into the dropper tube. Squirt this white gas across the top of the flame (Figure 2-10). It should ignite and resemble a small flame thrower. What were these white vapors? To confirm your conclusion, blow out the candle flame again, and quickly try to ignite the white vapors coming from the wick. You should see a flame propagate down this vapor trail and anchor on the wick. This propagating flame is a premixed flame; the anchored flame is, of course, our diffusion flame.

Review all of the information deduced about the candle flame and you should have a good understanding of the diffusion flame and fire. You may wish to consult Faraday's text for more informative experiments with the candle.[1]

Anatomy of a Diffusion Flame

More sophisticated experiments with a laminar diffusion flame using methane (CH_4) as the fuel have been conducted by Kermit Smyth and coworkers.[2,3] Their flame is formed from controlled flows of CH_4 and air, as shown in Figure 2-11. The burner is a slot burner, so it is planar in character compared to the cylindrical character of the candle, but the flame dynamics are the same. Various point measurement techniques were used in the flame to quantitatively reveal its anatomy. We can relate these measurements to what we have deduced from the candle experiments. Now our qualitative observations can be quantitatively presented. These quantitative results will be displayed for a height 9 mm above the burner.

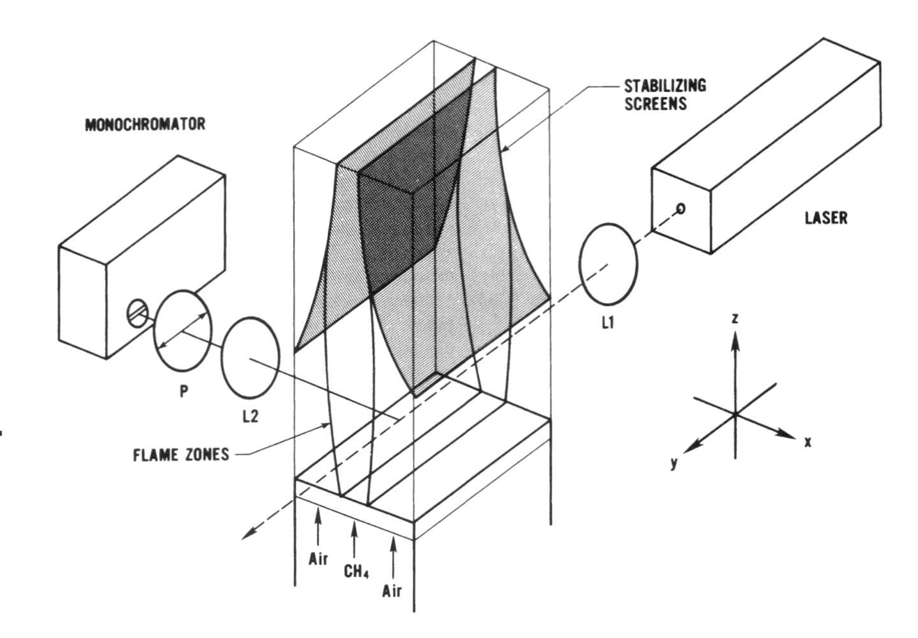

Figure 2-11 *Arrangement of slot burner flame and point measurement techniques. From Smyth et al., Ref. 2.*

thermocouple
device made of two dissimilar metal wires to measure temperature

Thermocouple measurements of temperature across the flame region are shown in Figure 2-12 at the 9-mm height. Other heights within the luminous flame region will be similar. These 9-mm measurements are indicative of the two sides of the diffusion flame. It is difficult to accurately measure the highest temperatures indicative of the reaction zone because the thermocouple probe will lose some heat and slightly cool the flame. Therefore, the high temperatures of approximately 1,700°C are likely to be more like 1,900° to 2,000°C. Nevertheless, this very high temperature region is indicative of the chemical reaction zone of the flame. This was the luminous yellow shell of the candle flame. The interior temperatures are due to heat conducted inward. The outer-wing temperature should eventually become the temperature of the pure air.

The corresponding vertical flow speed is shown in Figure 2-13. These flow speeds are principally due to buoyancy.

radicals
short-lived unstable molecules such as OH

soot
carbonaceous particles produced in flames

Using special laser-optical techniques Smyth and coworkers were able to image the visible flame, OH **radicals**, and **soot**. The OH radicals are indicative of short-lived chemical species in the heart of the combustion reaction. They extend to the air side of the flame. The soot, or black smoke of the candle experiments, comes from within the (luminous) flame zone. This is clearly seen in their images illustrated in Figure 2-14.

cracking
pyrolysis; breaking gaseous molecules into other molecules

The soot is formed on the fuel side of the diffusion flame. It results from a complex process as the original fuel (CH_4) is heated on diffusing toward the flame. This intense heating results in **cracking** the CH_4 molecules into many other hydrocarbon molecules. As acetylene (C_2H_2) and other precursors form, soot (mainly C atoms) is produced. The soot (a solid formed from gases) migrates through the reac-

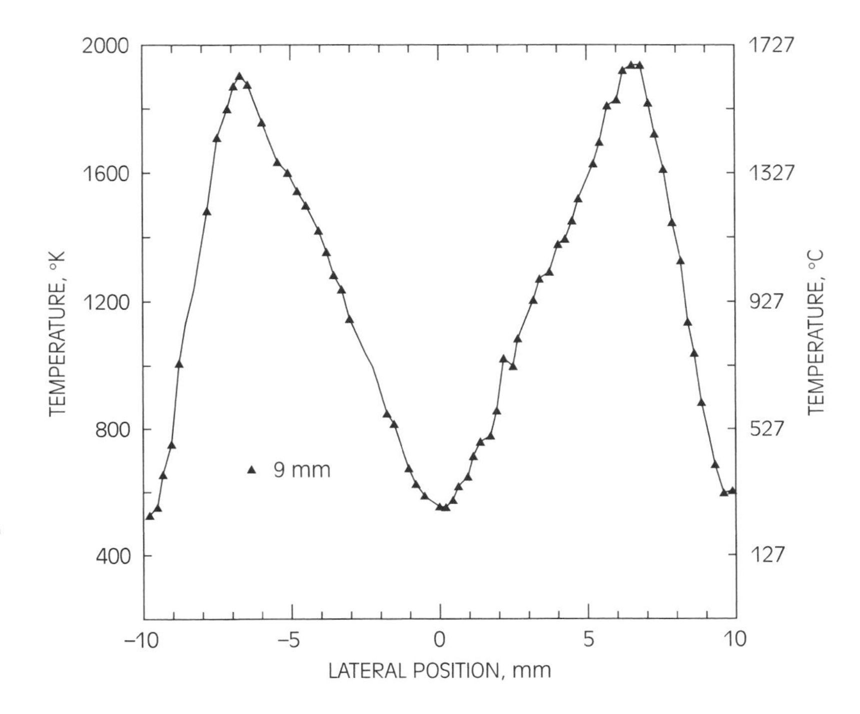

Figure 2-12 *Temperatures across a laminar diffusion flame. From Smyth et al., Ref. 2.*

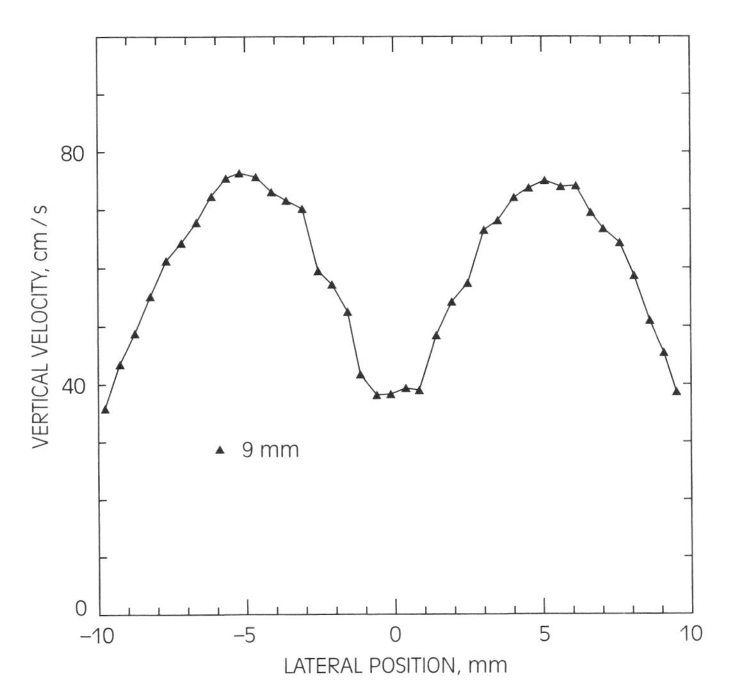

Figure 2-13 *Vertical velocity in a laminar diffusion flame. From Smyth et al., Ref. 2.*

tion (flame) zone, is oxidized (mainly by OH), and is consumed. We see this oxidation by the yellow (incandescent) glow of the visible flame. In this flame, as well as the candle flame, nearly all the soot is consumed before the reaction zone ends. The visible flame height nearly coincides with the end of the soot height (Figure 2-14) and the OH zone is slightly above this height. If we "tear" the flame by "stretching" it (e.g., abruptly pull the candle flame—be careful not to spill melted wax), soot can escape through the tear. Again, this illustrates that the soot formation is on the fuel side of the flame. For a heavily sooting fuel, all the soot may not be oxidized before the end of the reaction zone; then soot escapes as black smoke.

Let us examine the other compounds that occur in this flame process. Some of these are shown in Figure 2-15: The measurements were taken by Smyth et al.[2] at the same 9 mm above the fuel port. The concentrations are given in mole fraction, which is the same as the fraction by volume. For example, extract the oxygen at some point, let it come into balance with the air pressure; the ratio of its equilibrated volume to the volume of the mixture of gas is the same as the mole fraction. Note the oxygen (O_2) in pure air is at 0.21 with the nitrogen (N_2) making up the remainder, 0.79. These values are measured at the extreme left and right wings in Figure 2-15. The fuel (methane, CH_4) is nearly 1 (all fuel) at the center. We see that the fuel and oxygen reach zero at approximately 6 mm, the peak of the combustion zone. It is probably less than 1 mm in width at a height of 9 mm above the fuel source. The water vapor (H_2O) is also at maximum here. Along with carbon dioxide (CO_2), it is a principal compound formed and released into the atmosphere as a product of combustion. Figure 2-16 shows the results for CO_2 along with

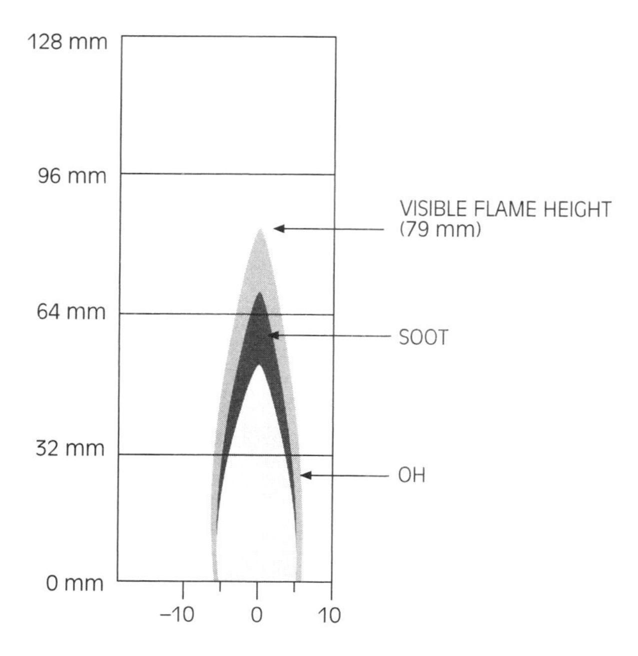

Figure 2-14 *Soot and flame (reaction) zone locations in a laminar diffusion flame. After Smyth et al., Ref. 3.*

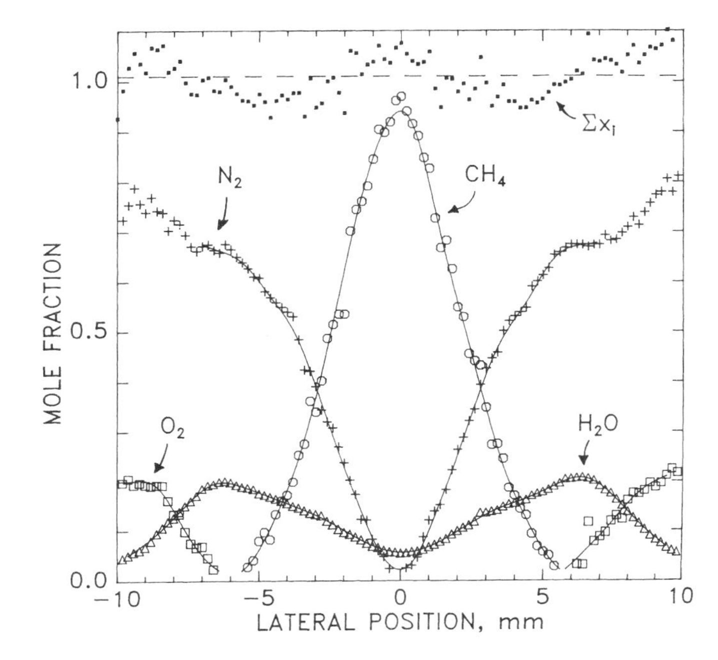

Figure 2-15 *Major species formed in a laminar diffusion flame (CH_4) at 9mm. From Smyth et al., Ref. 2.*

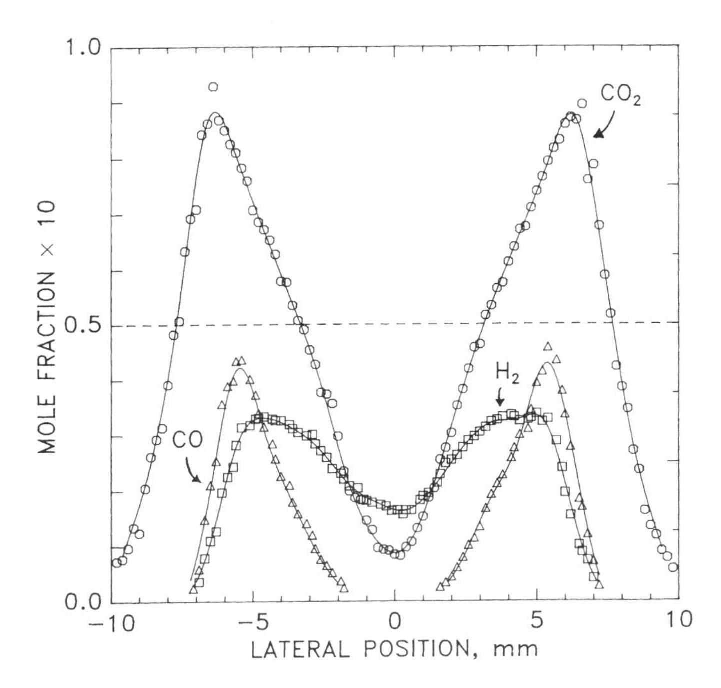

Figure 2-16 *Additional major species formed at 9 mm in a CH_4 laminar diffusion flame. From Smyth et al., Ref. 2.*

other compounds, namely, carbon monoxide (CO) and hydrogen (H_2), which do not survive to be released into the atmosphere.

The concentration of these species vary laterally because their relative proportions change due to the destruction of species. Figure 2-17 qualitatively illustrates these processes and some of the complex chemical processes taking place. Superimposed on these processes are the physical processes involving diffusion of species, each migrating toward its lowest concentration and involving buoyant flow due to the high temperatures.

synthesis recombination of molecules

Incomplete products of combustion species developed from incomplete combustion, for example, CO, H_2

Due to cracking and **synthesis** of the CH_4, as the heat from the combustion process is conducted to the diffusing CH_4 molecules, other hydrocarbons, some of which are shown in Figure 2-18, are formed in reactions preliminary to combustion. These compounds are "minor species," achieving short-lived concentrations of 0.005 mole fraction at most before being completely oxidized. Again, if the flame is "stretched" or all of these cannot be oxidized (due perhaps to retardants added to the fuel or excess fuel relative to air), they will find their way into the atmosphere as **incomplete products of combustion**.

■ **NOTE** **Nearly all natural diffusion flames arising due to accidental fires will be turbulent.**

Turbulent Diffusion Flames

We have seen that in the candle flame flow its plume becomes turbulent in less than 1 ft of height. This is typical of fire under normal buoyancy conditions. Therefore, nearly all natural diffusion flames arising due to accidental fires are turbulent. The laminar flame processes we have just reviewed must be viewed as a

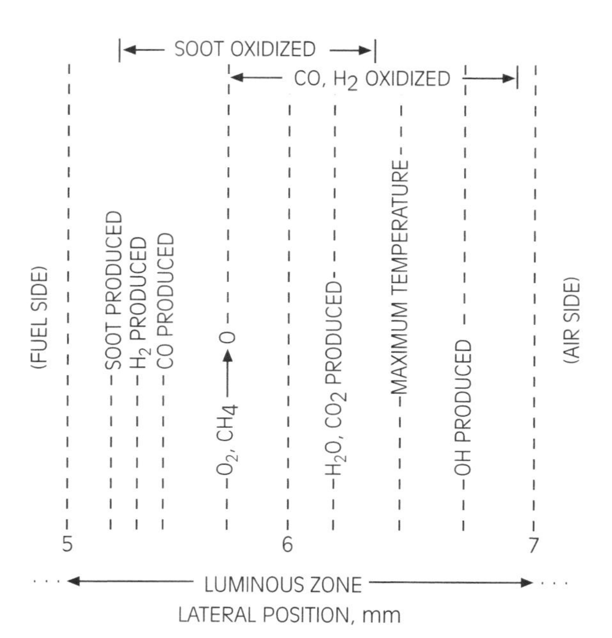

Figure 2-17 *Qualitative chemical processes in CH_4 laminar diffusion flame.*

fluctuating erratic processes. The processes depicted in Figures 2-12 to 2-18 are pushed and pulled in space, blurring the values of the concentrations. This blurring reduces their average peak values. For example, an attempt to measure temperature at a point where we think the flame is located produces a "flapping" flame zone. Consequently, the measurement probe will experience high and low temperatures, with averages approximately 800° to 1,000°C (see Figure 2-19).

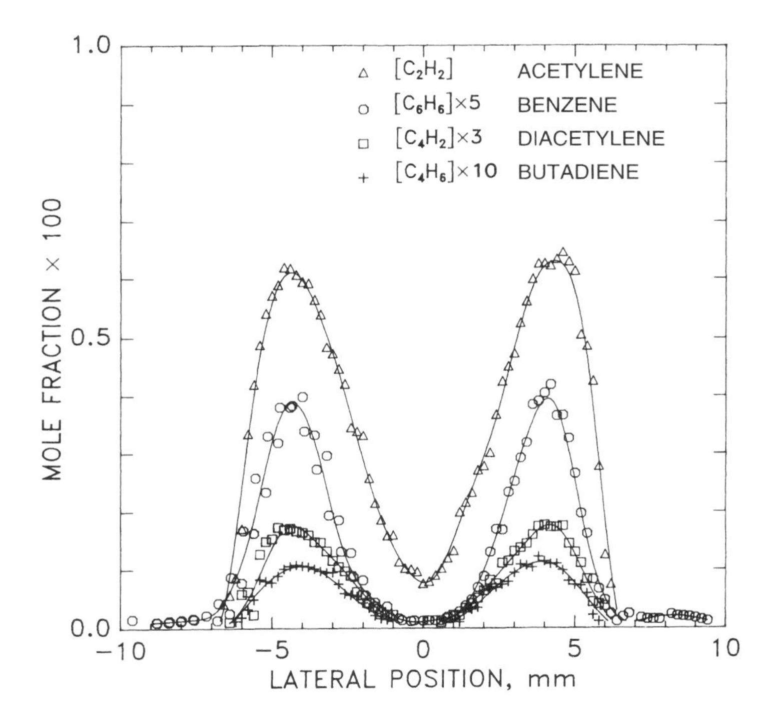

Figure 2-18 *Minor species in the CH_4 laminar diffusion flame at 9mm. From Smyth et al., Ref. 2.*

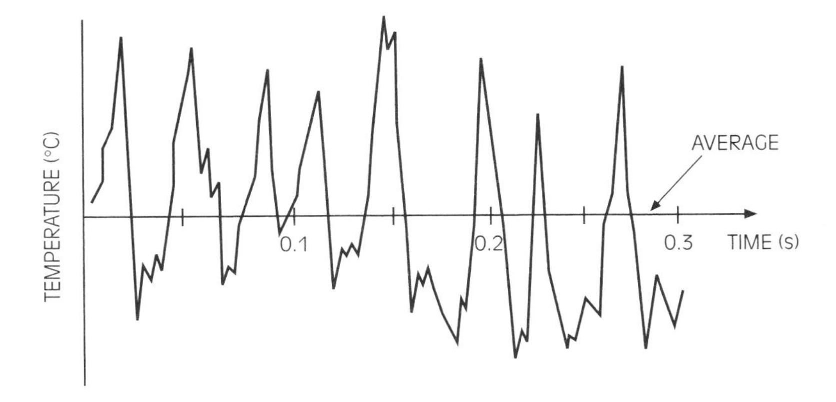

Figure 2-19 *Turbulent temperature fluctuations in a flame.*

SMOLDERING

smoldering
a slow combustion process between oxygen and a solid fuel

charring
the production of a solid charbonaceous residue on heating or burning a solid

Smoldering is a relatively slow combustion process that occurs between oxygen in the air and a solid fuel. The reaction occurs on the solid surface, and oxygen diffuses to the surface. The surface undergoes glowing and **charring**. The glowing is indicative of a temperature in excess of 1,000°C. Smoldering can revert to flaming combustion particularly due to a change, typically an increase, in the air flow rate over the fuel. Upholstered furniture and mattresses initiated by a discarded cigarette are prevalent sources of smoldering fires in homes. The incompleteness of the combustion process leads to high levels of carbon monoxide, CO, instead of CO_2. More than 10% of the fuel mass is converted to CO. Common examples of smoldering are shown in Figure 2-20.

■ **NOTE**
Glowing is indicative of a temperature in excess of 1000°C.

■ **NOTE**
During incomplete combustion, more than 10% of the fuel mass is converted to CO.

■ **NOTE**
Smoldering is very slow, but potentially deadly due to its production of CO.

Smoldering can occur in porous solid fuels, in combinations of fuels, in impervious solids, and in buried solid fuel waste sites. It requires air, but not much, because the process is so slow. Air can flow toward the smolder front in either direction and can be naturally induced to flow due to buoyancy. Smoldering is schematically illustrated in Figure 2-21. The speed of the smoldering reaction depends on many factors, but is generally of the order of 10^{-2} to 10^{-3} cm/s or roughly 1–5 mm/minute. Smoldering is very slow but potentially deadly because it produces carbon monoxide. Typical smolder velocities are shown in Figure 2-22 and Table 2-1. These apply to forward smolder (air moving in the same direction as the smolder front) and reverse smolder (air moving in the opposite direction to the smolder front).

CIGARETTE

UPHOLSTERED CHAIR
FOAM PLASTIC OR COTTON
AND FABRIC SYSTEM

CHARCOAL GRILL

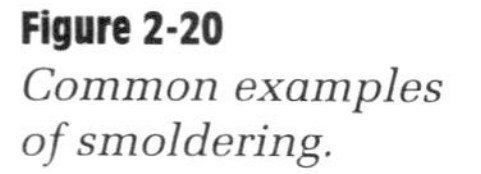

Figure 2-20
Common examples of smoldering.

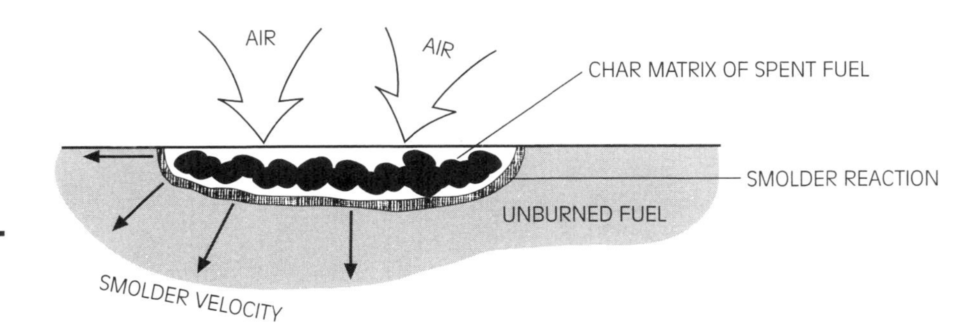

Figure 2-21 *Schematic of smoldering.*

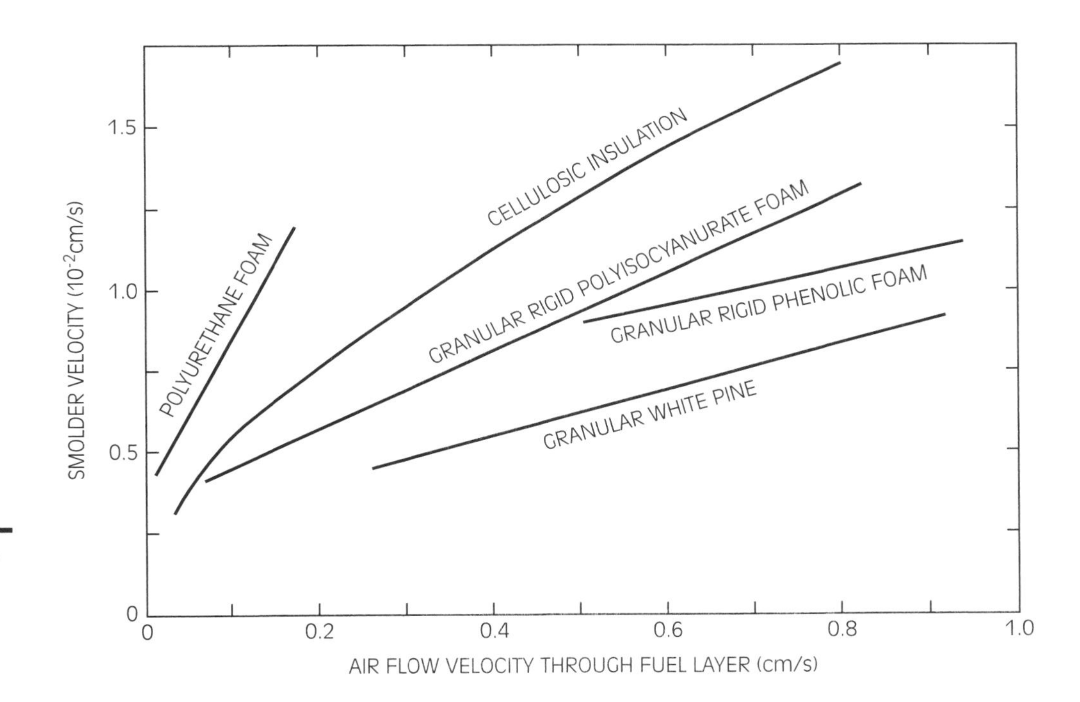

Figure 2-22 *Reverse smolder velocities dependent on air flow. After Ohlemiller, Ref. 4.*

SPONTANEOUS COMBUSTION

spontaneous combustion a process by which combustion occurs after a self-incubation period between a fuel and oxidizer

Spontaneous combustion is a combustion process that can begin with a slow oxidation in a fuel exposed to air. The chemical reaction is relatively slow so that a combustion process may not be perceptible or may at most be in a low smoldering state. The release of energy by the chemical reaction competes with the ability of the fuel to lose heat to the surrounding air. If insufficient heat is lost, the temperature of the fuel can increase, which in turn causes a faster chemical reac-

Table 2-1 *Typical smolder velocities associated with fuel configuration.*

Fuel	Fuel/Smolder Configuration	Air Supply Condition/Rate	Smolder Velocity (cm/sec)
Pressed fiber insulation board, 0.23–0.29 g/cc	1.3 cm thick, horizontal strips, width large compared to thickness	Natural convection/diffusion	$1.3–2.2 \times 10^{-3}$
Pressed fiber insulation board, 0.23–0.29 g/cc	1.3 cm × 1.3 cm strips varied angle to vertical	Natural convection/diffusion	$2.7–4.7 \times 10^{-3}$
Pressed fiber insulation board, 0.23–0.29 g/cc	1.3 cm × 5 cm strips forward smolder	Forced flow, 20 to 1500 cm/s	3.5×10^{-3} (in 20 cm/s air) 13.0×10^{-3} (in 1400 cm/s air)
Pressed fiber insulation board, 0.23–0.29 g/cc	1.3 cm × 5 cm strips reverse smolder	Forced flow, 80–700 cm/s	$2.8–3.5 \times 10^{-3}$
Pressed fiberboard (pine or aspen) 0.24 g/cc	1.3 cm × 30 cm sheets, horizontal, forward smolder	Forced flow, 10–18 cm/s	0.7×10^{-3}
Cardboard	Vertical rolled cardboard cylinder, downward propagation, varied dia. 0.19–0.38 cm	Natural convection, diffusion	$5.0–8.4 \times 10^{-3}$
Shredded tobacco	0.8 cm dia. cigarette, horizontal, in open air	Natural convection, diffusion	$3.0–5.0 \times 10^{-3}$
Cellulose fabric + 3% NaCl	Double fabric layer, 0.2 cm thick, horizontal, forward smolder	Forced flow, ≈ 10 cm/s	$\approx 1.0 \times 10^{-2}$

Source: After Ohlemiller, Ref. 4.

thermal runaway
an accelerating chemical reaction due to heat transfer

■ NOTE
The larger the fuel array, the more the system is prone to spontaneous combustion.

tion. This reaction leads to a **thermal runaway**, and eventually the process can result in flaming combustion (a diffusion flame) or active smoldering. Because heat loss is a factor, the larger the fuel array, the more the system is prone to spontaneous combustion. The entire process from inception to flaming may take hours or days. It requires a critical set of environmental or heating conditions to be possible. Although theories exist for spontaneous combustion (or ignition), they are difficult to apply with high accuracy. Some examples of different fuel media prone to spontaneous ignition are shown in Figure 2-23.

All chemical reactions leading to fire increase rapidly as the temperature increases. If the size of the fuel and air mixture inhibits the rate of heat loss, energy can build up to raise the temperature. This increased temperature has the potential to accelerate the reaction from one of yellowing cotton fabric over

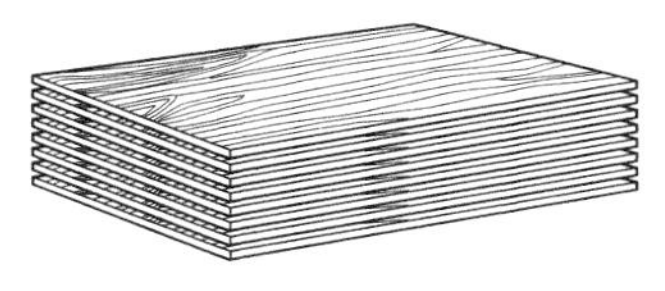

MIXTURE OF FUEL GAS AND AIR RAISED TO ITS AUTOIGNITION TEMPERATURE

HAYSTACK, BIOLOGICAL REACTIONS PROMOTED BY MOISTURE

PILE OF WOOD FIBERBOARD OR COAL WASTE (SLAG)

Figure 2-23 *Examples of fuel arrays prone to spontaneous ignition.*

decades to an ignition in a box of cotton waste in hours. An excerpt from Bowes[5] describes the range of results for cotton waste soaked with linseed oil:

> The amounts of material involved in the initiation of fires can be quite small. For example Taradoire [1925], investigating the self-ignition of cleaning rags impregnated with painting materials, stated that experience indicated the occurrence of self-ignition in as little as 25 g of rags but, in a series of experiments, found that the best results (sic) were obtained with 75 g of cotton rags impregnated with an equal weight of mixtures of linseed oil, turpentine and liquid "driers" (containing manganese resinate) and exposed to the air in cylindrical containers made of large-mesh wire gauze (dimensions not stated). Times to ignition generally lay between 1 h and 6 h but, for reasons which were not discovered, some samples, which had not ignited within these times and were considered safe, ignited after several days. Other examples giving an indication of scale have been described by Kissling [1895], who obtained self-heating to ignition, from an initial temperature of 23.5°C, in 50 g of cotton wool soaked with 100 g of linseed oil and by Gamble [1941] who reports two experiments with cotton waste soaked in boiled linseed oil (quantities not stated) and lightly packed into boxes; with the smaller of the two, a cardboard box of dimensions 10 cm × 10 cm × 15 cm, the temperature rose from 21°C to 226°C in 6 1/4 h and the cotton was charred.

The propensity to spontaneous ignition can be measured by subjecting a sample of the substance in question to oven air temperatures until the lowest air temperature is found to eventually (in hours or days) cause ignition, which can be manifested by smoldering or flaming. As the size of the fuel increases, reducing the ability of heat to be conducted from its center, the critical air temperature needed for ignition decreases. Additives, such as certain oils, can enhance this propensity. Bowes[5] reports such experimental results in Table 2-2. Such testing and its data can be used to assess the potential for spontaneous ignition, but because the

Table 2-2 *Sawdust cubes, with and without oil, exposed to air temperatures.*

Cube Size 2r mm	Oil Content (%)	Ignition Temperature (°C)
25.4	0	212
25.4	11.1	208
51	0	185
51	11.1	167
76	0	173
76	11.1	146
152	0	152
152	11.1	116
303	0	135
303	11.1	99
910	0	109
910	11.1	65

Source: From Bowes, Ref. 5.

premixed flame
a flame in which fuel and air are mixed first before combustion

shock wave
abrupt change in temperature and pressure due to a flow instability caused by speeds in excess of the speed of sound

detonation
a premixed flame preceded by a shock wave

upper and lower flammability limits
concentrations of fuel in air in which a premixed flame can propagate

NOTE

In a confined space, a premixed flame can cause a rapid pressure increase that manifests itself as an explosion.

real conditions may not be duplicated completely in the test oven, extrapolations are limited in accuracy.

PREMIXED FLAMES

A **premixed flame** is a combustion process in which the fuel gas and air (or oxygen) are first mixed before ignition and propagation occur. In a confined space, such a process can cause a rapid pressure increase that manifests itself as an explosion. If sufficient pressure builds up behind the propagating flame, a **shock wave** can form ahead of the flame. This event is then called a **detonation**. Not only does such an event have great damage potential, but the pressure-driven flame pushes the flame to enormous speeds. Examples of premixed flames are shown in Figure 2-24.

Returning to the candle experiments: When the flame is blown out, the white wax-fuel vapor mixes with air as it rises in its distinct plume. Attempts to reignite the plume will succeed provided the ignition flame is placed in the plume stream at the right concentration. Gaseous fuels will ignite within distinct fuel concentration limits. For every fuel gas in air, there is an **upper (UFL) and lower (LFL) flammable limit concentration** in which flame propagation can occur. These values (usually calculated at 25°C in air at normal pressure) vary slightly with the

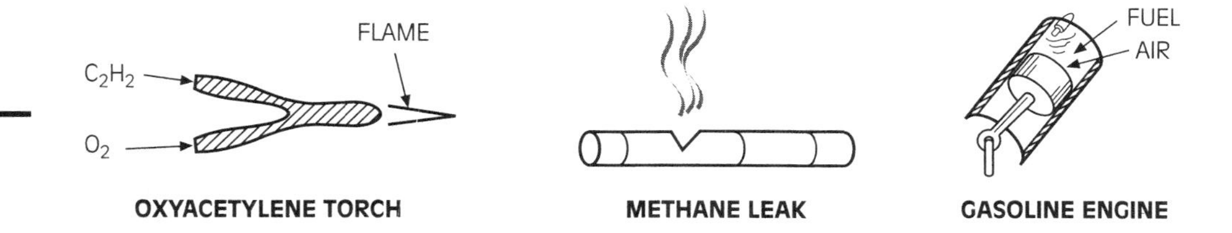

Figure 2-24 *Examples of premixed flames.*

flashpoint temperature of a liquid fuel needed to cause ignition with a heat source; corresponds to the lower flammable limit of the evaporating vapor

autoignition temperature the lowest temperature at which a mixture of fuel and oxidizer can propagate a flame with no other heating

temperature as shown in Figure 2-25. At a given temperature we see there is an upper and lower flammability limit. In this range, this means a small energy source will initiate sustained propagation at a speed that would have been seen for the candle vapor ignition. It is laminar and generally less than 1 m/s. If the temperature of the fuel mixture is varied, the LFL and UFL change slightly as seen in Figure 2-25. But at very low temperatures, the fuel may be condensed and in the form of droplets. This droplet mixture also has flammability limits. The temperature (T_L) just before the fuel condenses to a liquid at the LFL is called the **flashpoint**. We discuss this phenomenon further when we examine the ignition of liquids in Chapter 4. Although this ignition process requires an energy source such as a flame or electric discharge, it is quite small—less than 1 mJ (which is that amount of energy to raise 1 gram of water less than 0.001°C). In contrast, at a high enough temperature, the fuel–air mixture can ignite without any energy source. The lowest temperature to cause this spontaneous ignition phenomenon is called the **autoignition temperature** (AIT), also illustrated in Figure 2-25.

Values for the LFL and UFL in mole (or volume) concentrations in percentages are shown in Table 2-3. The autoignition temperatures are also listed.

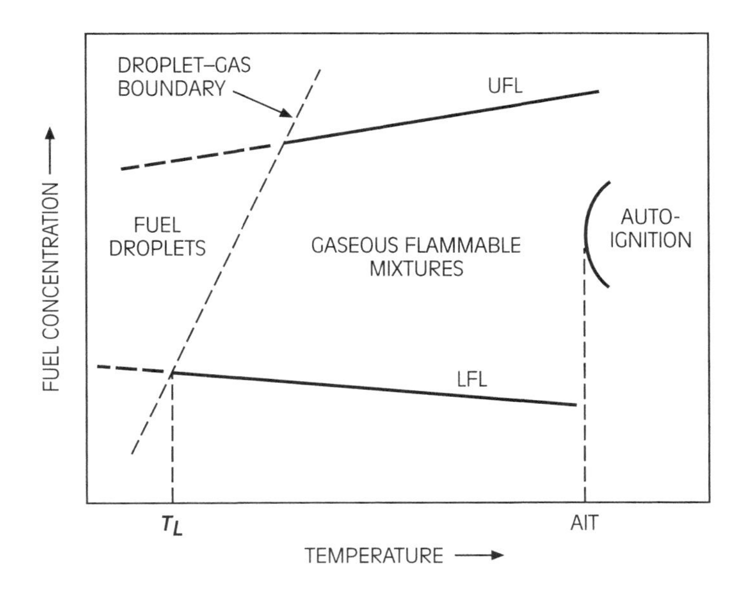

Figure 2-25 *Limits of flammability in air showing effect of temperature. After Zabetakis, Ref. 6.*

Table 2-3 *Flammability limits of gaseous fuels in air at normal atmospheric temperature and pressure.*

	LFL (%)	UFL (%)	AIT (°C)
Acetylene	2.5	100	305
Benzene	1.3	7.9	560
n-Butane	1.8	8.4	405
Carbon Monoxide	12.5	74	609
Ethylene	2.7	36	490
n-Heptane	1.05	6.7	215
Hydrogen	4.0	75	400
Methane	5.0	15.0	540
Propane	2.1	9.5	450
Trichlorethylene	12	40	420

Source: Based on data from Beyler, Ref. 7.

Once propagation commences in the fuel–air mixtures from the ignition point, flame propagation occurs under laminar conditions at speeds of roughly 0.1 to 0.5 m/s. As turbulence occurs, the "wrinkled" flame will have more surface area and hence more energy so its speed will increase. In the extreme case of a pressure-driven flame leading to detonation, the velocity accelerates to over 2,000 m/s. Limits of detonation have been determined and these are listed in Table 2-4 for some fuel mixtures along with the detonation flame velocities. The detonation limits appear to be within the normal flammability limits (Table 2-4).

Table 2-4 *Detonation limits at normal atmospheric temperature and pressure.*

	Lower Limit (%)	Upper Limit (%)	Detonation Velocity (m/s)
Hydrogen in pure O_2	15	90	2821
Hydrogen in air	18.3	59	—
CO in pure moist O_2	38	90	1264
Propane in pure O_2	3.2	37	2280, 2600
Acetylene in air	4.2	50	—
Acetylene in pure O_2	3.5	92	2716

Source: From Lewis and von Elbe, Ref. 8.

Summary

Natural fire processes can be divided into four forms of combustion:

1. Spontaneous combustion,
2. Smoldering,
3. Premixed flames,
4. Diffusion flames.

Spontaneous combustion, or autoignition, occurs because the fuel undergoes an exothermic reaction (generating chemical energy) that cannot be cooled. As a result, the reaction accelerates and leads to either flaming or smoldering combustion. Spontaneous combustion can result from biological decay, as in moist haystacks, or by oxidation, as in the drying of linseed oil on cotton. It can take hours or days to develop perceptible combustion.

Smoldering is the oxidation process with solids that results in temperatures of 400–1,000°C and propagation rates of 10^{-3} to 10^{-2} cm/s.

Premixed flames are the vehicle for piloted ignition as seen in igniting the white vapors of the blown out candle flame. Such propagation requires specific fuel concentrations within the upper and lower flammable limits and commences at speeds of about 30 cm/s, but can accelerate to speeds greater than the speed of sound as the result of a detonation.

Diffusion flames form the basis of flaming uncontrolled fires in structures and wildland. Understanding the nature of the processes in bringing the fuel gases and air to the flame zone, as illustrated by the candle flame experiments, can give a good concept for the nature of diffusion flames. Fuel and oxygen are consumed in the flame, and soot is produced on the fuel side of the flame. Also, the process can be viewed equally as burning all the fuel or all the oxygen.

Review Questions

1. Conduct and discuss the candle experiments described in this chapter.
2. Measure the velocity of a smoldering process, e.g., a cigarette. Explain the process.

True or False

1. Autoignition requires a pilot flame.
2. An oxyacetylene torch is a diffusion flame.
3. A burning pool of gasoline is a diffusion flame.
4. Charring is not likely during smoldering.
5. Plastics cannot smolder.
6. Smoldering can occur in concealed spaces with very little air.
7. A flame must be 10 ft high to be turbulent.

Activities

1. Put linseed oil soaked (not too wet) cotton rags in a cardboard box about 1 ft^3. Place in a safe place for doing fire tests and let sit for several hours. Explain the results.
2. Compare the different flames produced by a Bunsen burner by regulating the air vents. Can you blow the flame off? Is it "swallowed" into the tube?

References

1. Michael Faraday, *Faraday's Chemical History of the Candle* (Chicago, IL: Chicago Review Press, 1988 [orig. 1861]).
2. Kermit C. Smyth, J. Houston Miller, R. C. Dorfman, W. G. Mallard, and Robert J. Santoro, "Soot Inception in a Methane/Air Diffusion Flame as Characterized by Detailed Species Profiles," *Combustion and Flame* 62 (1985): 157–181.
3. K. Smyth, J. E. Harrington, E. L. Johnson, and W. M. Pitts, "Greatly Enhanced Soot Scattering in Flickering CH_4/Air Diffusion Flames," *Combustion and Flame*, 95, 1993: 229–239.
4. T. J. Ohlemiller, "Smoldering Combustion," chap. 2-11 in *SFPE Handbook of Fire Protection Engineering*, 2d ed., edited by P. J. DiNenno (Quincy, MA: National Fire Protection Association, June 1995).
5. P. C. Bowes, *Self-Heating: Evaluating and Controlling the Hazards* (London: Her Majesty's Stationery Office, 1984).
6. M. G. Zabetakis, *Flammability Characteristics of Combustible Gases and Vapors*, Bureau of Mines, Bulletin 627 (Washington, DC: U.S. Department Of the Interior, 1965).
7. C. L. Beyler, "Flammability Limits of Premixed and Diffusion Flames," chap. 2-9 in *SFPE Handbook of Fire Protection Engineering*, 2d ed., edited by P. J. DiNenno, (Quincy, MA: National Fire Protection Association, June 1995).
8. B. J. Lewis and G. von Elbe, *Combustion, Flames and Explosions of Gases*, 3d ed. (Orlando, FL: Academic Press, 1987), 551–555.

Heat Transfer

Learning Objectives

Upon completion of this chapter, you should be able to:

- Understand the difference between energy and heat.
- Explain conduction, convection, and radiation heat transfer.
- Describe the concept of heat flux (e.g., kW/m^2) and its significance to hazards of heat transfer from fire.
- Perform simple computations in heat transfer.

INTRODUCTION

Although chemistry and fluid mechanics are important in the study of combustion, we can forgo their study for elementary quantitative purposes. However, to develop the ability to make quantitative analyses for fire we must first develop some understanding and computational ability in heat transfer.

DEFINITIONS AND CONCEPTS

energy
a state of matter representative of its ability to do work or transfer heat

work
the movement of mass over a distance

thermal energy
energy directly related to the temperature of an object

Rankine (°R)
absolute Fahrenheit temperature scale, 460 + °F

Kelvin (K)
absolute Celsius temperature scale, 273 + °C

The study of **energy** is rooted in the subject of thermodynamics, a very logical science that carefully defines energy, heat, temperature, and other properties. As with fluid flow, heat transfer, and chemistry, thermodynamics is at least a semester course in itself. Therefore, we must treat certain concepts in little depth.

Energy can take many forms. It cannot be destroyed (unless we consider nuclear reactions), but it can assume different forms. Energy due to motion is kinetic, due to flowing charges is electrical, and due to temperature is thermal. In the study of fire, a key form of energy is chemical. Energy is a state of matter from which we might extract **work** or obtain heat. For example, in a chemical reaction, thermal (internal) energy is converted to chemical energy, which in a fire gives rise to heat transfer. In an automobile, chemical energy is converted into work that drives the wheels and heat that is lost to the air from the engine block.

Thermal (or internal) energy is a property of matter directly associated with the concept of temperature. Something that is "hot" has a relatively higher internal energy and temperature compared to that which is "cold." The scales used to measure temperature are arbitrary and have been established by convenience as well as by science. The relationship of four scales is illustrated in Figure 3-1. (See Table 1-5.) They have been based on the freezing and boiling points of water. Two absolute scales, **Rankine** (°R) and **Kelvin** (°K), set zero temperature at zero inter-

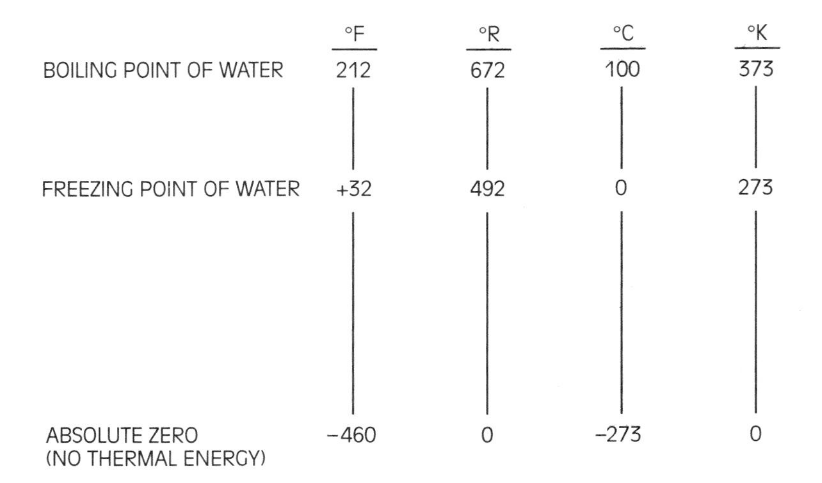

Figure 3-1
Temperature scales.

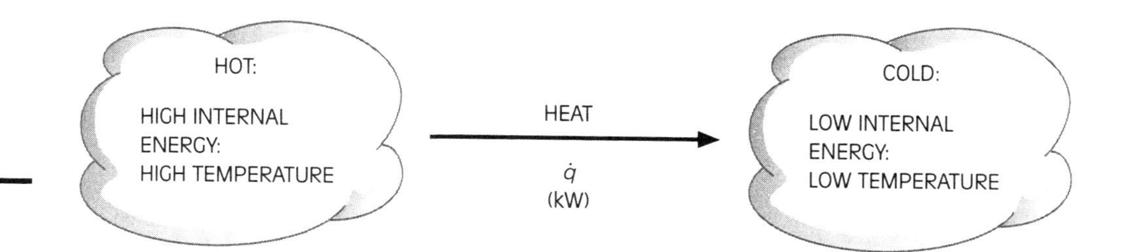

Figure 3-2 *Heat transfer process.*

heat
energy transfer due to temperature difference

conduction
heat transfer due to molecular energy transfer following Fourier's Law

radiation
heat transfer due to electromagnetic energy transfer such as light

nal energy. At this state, the molecular mechanical activity ceases. Temperature is a direct measure of this activity.

Heat is thermal energy in motion that travels from a "hot" to a "cold" region. There are two laws that govern this heat transfer: One needs matter to travel within, the other allows travel of heat in a vacuum. This process is symbolically illustrated in Figure 3-2.

The symbol q represents heat; Q represents energy. Both are measured in joules (J). It takes 4.182 J to raise 1 gram (g) of water 1°C. (See Tables 1-4 and 1-5. Additional notation is found in Table 1-5.) The rate of heat flow or the rate of energy generated (transformed) is given in kJ/s or kW, kilowatts. Hopefully, before the end of this book, you will relate to fire in terms of its kW output in the same way you relate to ratings for a light bulb, furnace, or toaster.

FORMS OF HEAT TRANSFER

■ NOTE
Fourier, in the early 1800s, formulated the law of heat conduction.

■ NOTE
Planck's quantum theory led to a theoretical basis for radiation heat transfer.

The laws of **conduction** and **radiation** govern heat transfer. In the early 1800s, Joseph Fourier formulated the law of heat conduction, which states that the rate of heat flow through matter is directly proportional to temperature difference. Formulated in the early 1900s, Max Planck's quantum theory led to a theoretical basis for radiation heat transfer. Such radiation moves at the speed of light in a vacuum.

Convection is another category of heat transfer that is a subset of conduction. It applies specifically to conduction in a moving fluid. Figure 3-3 illustrates these forms of heat transfer in a fire situation of a flame against a wall. Conduction occurs through the wall, convection occurs at the wall–flame interface, and radiation can be felt at a distance.

convection
conduction heat transfer from a moving fluid (gas or liquid) to a solid surface

Conduction

The simplest representation of the law of conduction is the steady (unchanging with time) flow of heat through a wall as illustrated in Figure 3-4. The equation is given below:

$$\dot{q} = kA(T_2 - T_1)/l \quad \textbf{(3-1)}$$

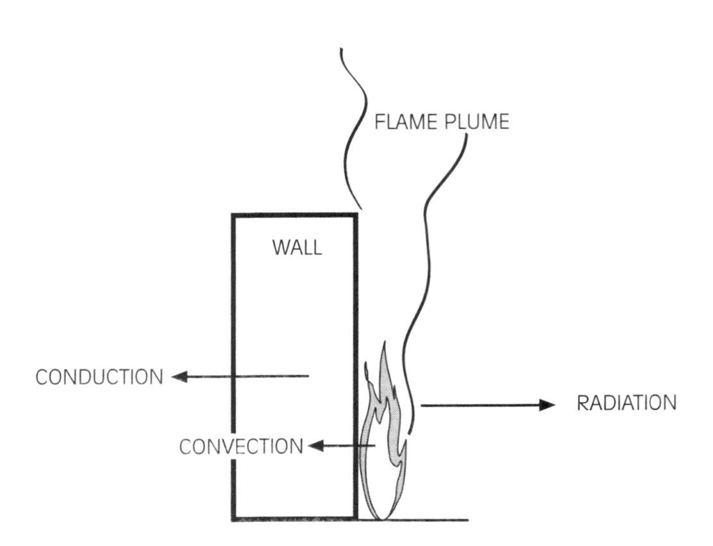

Figure 3-3 *Examples of heat transfer in fire. Conduction: heat transfer in a stationary medium; convection: heat transfer by conduction in a moving fluid to a solid; radiation: heat transfer by electromagnetic energy generated due to an object having a temperature.*

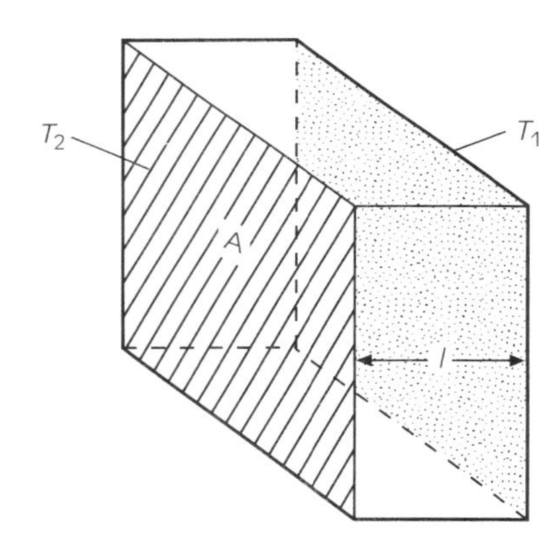

Figure 3-4 *Heat conduction through a wall.*

thermal conductivity the property of matter that represents the ability to transfer heat by conduction

where k is the **thermal conductivity**,
A is the area through which the heat is transferred,
T_2 and T_1 are the respective temperatures of the wall faces, and
l is the wall thickness.

This equation says that the rate of heat flow (e.g., kJ per second or kW) between two temperatures, T_1 and T_2, in a solid is proportional to a property of the solid known as thermal conductivity (k). Also, the larger the flow area along which the heat is conducted, the greater is the flow rate, $\dot{q}$. This is analogous to water flowing through a pipe. The commonly used R-value for home siding insulation is the reciprocal of kA/l. This reciprocal is called **thermal resistance** as an analogy to electrical resistance where $\dot{q}$ takes on the role of current flow rate and temperature takes the role of voltage. Let us consider an example.

thermal resistance
the inverse of conductivity times area per unit length (l/kA), or the temperature difference required per rate of heat conducted

heat flux
heat flow rate per unit area of flow path

Example: Find the heat flow rate per unit area through a wall of polyurethane foam, 0.05 m thick(l), across a temperature difference of 40°C to 20°C. Note that heat flow rate per unit area is called **heat flux** (denoted by $\dot{q}''$).

$$\dot{q}'' = \dot{q}/A = (0.034 \text{ W/m-K})(40 - 20)\ ^\circ\text{C}/0.05 \text{ m} = 13.6 \text{ W/m}^2,$$
$$\text{or } 0.013 \text{ kW/m}^2$$

Consider the same conditions, except the solid is now steel. Here, k (from Table 3-1) is 45.8 W/m·K. Note that whether we use K or °C in the thermal conductivity term (or for the temperature in the formula), there is no change, because we are concerned with temperature differences. The calculation for this case is

Table 3-1 *Table of thermal properties.*

Material	Thermal Conductivity (k) (W/m-K)	Specific Heat (c) (kJ/kg-K)	Density (ρ) (kg/m³)	Thermal Diffusivity (α) (m²/s)	Thermal Inertia ($k\rho c$) (kW²-s/m⁴-K²)
Copper	387	0.380	8940	1.14×10^{-4}	1300.0
Steel (mild)	45.8	0.460	7850	1.26×10^{-5}	160.0
Brick (common)	0.69	0.840	1600	5.2×10^{-7}	0.93
Concrete	0.8–1.4	0.880	1900–2300	5.7×10^{-7}	2.0
Glass (plate)	0.76	0.840	2700	3.3×10^{-7}	1.7
Gypsum plaster	0.48	0.840	1440	4.1×10^{-7}	0.58
PMMA	0.19	1.420	1190	1.1×10^{-7}	0.32
Oak	0.17	2.380	800	8.9×10^{-8}	0.32
Yellow pine	0.14	2.850	640	8.3×10^{-8}	0.25
Asbestos	0.15	1.050	577	2.5×10^{-7}	0.091
Fiber insulating board	0.041	2.090	229	8.6×10^{-8}	0.020
Polyurethane foam	0.034	1.400	20	1.2×10^{-6}	9.5×10^{-4}
Air	0.026	1.040	1.1	2.2×10^{-5}	3.0×10^{-5}

Source: From Drysdale, Ref. 1.

$$\dot{q}'' = \frac{(45.8 \text{ W/m-K})(20°\text{C})}{(0.05 \text{ m})} = 18{,}320 \text{ W/m}^2$$

or

$$\dot{q}'' = 18.3 \text{kW/m}^2$$

Thus, we see the ability of heat to penetrate steel through an insulator like polyurethane.

■ NOTE
Layers of different materials, their shape, and the time for thermal penetration all add to the complexity of heat transfer.

Under fire conditions, heat conduction can play a role in propagating the fire. Fires have been known to propagate on ships due to direct heat conduction through the steel floors and bulkheads. In general, the transfer of heat is more complex than we have depicted it by Equation (3-1). Layers of different materials, their shape, and the time for thermal penetration all add to the complexity.

In general, heat transfer is unsteady, and it takes some time for the heat to penetrate through the wall. As a rough estimate of how long it will take the back of the wall to feel that there is an increase in temperature on the front face, we use the following formula. This gives the time for the heat wave to penetrate the wall.

$$\text{Thermal penetration time} \approx \frac{l^2}{16\alpha}$$

where $\alpha = k/\rho c$, thermal diffusivity,
ρ is density, and
c is specific heat.

How long would it take the back wall in the previous polyurethane example to realize that heat has been applied to the front wall? The thermal diffusivity can be found from Table 3-1

$$\text{time} = (0.05 \text{ m})^2/(16(1.2 \times 10^{-6} \text{ m}^2/\text{s})) = 130 \text{ s}$$

In contrast, the steel wall has a penetration time of 12.4 seconds. Very roughly, steady conduction through the wall will take place after this penetration time. Then Equation (3-1) can be applied. All of these facets must be considered in analyzing heat conduction.

Convection

In a moving fluid, the heat transfer from the fluid to a solid surface is called convection but the governing law is conduction. However, the temperature difference and distance traversed near the surface is not directly measured. For example, as hot air flows over frozen water, the heat transfer by conduction at the ice surface depends on the temperature difference near the surface, ΔT. By the law of conduction, the heat transfer at the ice surface from the air as illustrated in Figure 3-5, is still given by Equation (3-1):

$$\dot{q} = kA\ \Delta T/l$$

where l is the distance between the temperatures corresponding to ΔT. The heat flux to the wall is

$$\dot{q}'' = \dot{q}/A = k\ \Delta T/l$$

Since ΔT and l are not available without detailed temperature measurements, the equation is rearranged as

$$\dot{q}'' = (k/l)(T_2 - T_1)$$

where T_2 is the air stream temperature (e.g., 30°C) and T_1 is the surface temperature (e.g., 0°C). The flow speed of the air will affect the distance l; the higher the flow speed, the more l will be reduced. The quantity (k/l) is defined as the **convective heat transfer coefficient** given by the symbol, h. It depends on the air properties as well as the flow speed. Therefore, the equation for a flowing fluid that gives the heat flux at the surface is given as

convective heat transfer coefficient
a quality that represents the ability of heat to be transformed from a moving fluid to a solid surface

$$\dot{q}'' = h\ (T_2 - T_1) \quad \textbf{(3-2)}$$

It is too complicated for us to explain all the methods for determining h. For fire processes, we do not need much information to make estimates. Table 3-2 gives typical values for h. Fire conditions do not significantly change the values given for air as suggested in Table 3-2. Let us consider a fire example.

Example: Find the convective heat flux from a turbulent flame to a cold wall at 20°C. Estimate the h value from the table for free convection (a buoyant flow that is indicative of a flame) as 5 W/m^2-°C. The maximum time-averaged flame temperature is approximately 800°C.

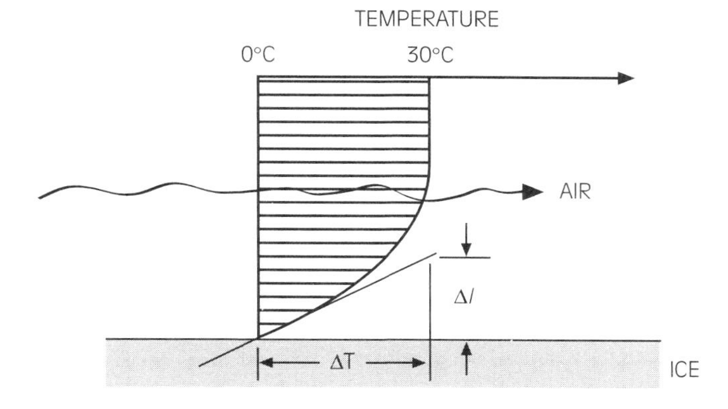

Figure 3-5
Convective heat transfer example.

Table 3-2 *Typical values for convective coefficients,* h

Fluid Condition	*h* (W/m^2°C)
Buoyant flows in air	5–10
Laminar match flame	~ 30
Turbulent liquid pool fire surface	~ 20
Fire plume impinging on a ceiling	5–50
2 m/s wind speed in air	~ 10
35 m/s wind speed in air	~ 75

■ **NOTE**
Computing radient heat flux to a target is very important for assessing potential damage and the possibility of a remote ignition.

$$\dot{q}'' = (5\ W/m^2\text{-}°C)\ (800 - 20)°C$$
$$= 3900\ W/m^2$$
$$= 3.9\ kW/m^2$$

Had we used 10 W/m^2-°C for h instead, our answer would be double. This range of 5 to 10 kW/m^2 is representative of most turbulent fire conditions. The convection heat flux to a burning (vaporizing) surface will be slightly less because the flow of fuel gases will push the hot flame away from the surface, increasing l (see Figure 3-5 on previous page). Also, for laminar flames, the flame will be closer to the surface, which corresponds to a smaller l. For this reason small laminar flames have a higher convective heat flux than turbulent flames.

Radiation

frequency
cycles per unit time (measured in hertz, cycles per second)

wavelength
the distance traveled in one cycle, or the speed of light divided by frequency

Planck established the theoretical basis for explaining radiation heat transfer. All radiation is electromagnetic energy consisting of both electric and magnetic fields. These radiation rays have both a **frequency** and a speed (the speed of light in a vacuum). The frequency range determines the descriptive name or phenomena we associate with the radiation. For example, high frequency (alternative low **wavelength**) radiation phenomena in descending order are cosmic rays, gamma rays, and x-rays reaching into ultraviolet, which borders visible light (radiation). In decreasing order from visible light, we move to infrared and then radio waves. Thermal radiation extends across the infrared and somewhat into the visible. Objects having temperatures of 1,000°C can appear bright red; we "see" the thermal radiation within the frequency or wavelength range of our eye. The use of an infrared camera extends our ability to see lower temperatures. Night vision cameras do this, making humans visible on a dark night.

Radiation originates in different ways. Radio waves can be electronically developed for broadcasting. For example at 100 MHz (Megahertz) or 100 million cycles per second on our FM dial we might find WFIR. Thermal radiation originates solely due to temperature. All matter having a temperature above absolute zero emits radiation. This radiation has a frequency range and can be predicted from Planck's theory, however, we will not examine radiation in that depth. Rather than examining how thermal radiation spans the radio dial we simply examine the overall output. The maximum possible output of radiation due to temperature is expressed in terms of heat flux as

$$\dot{q}'' = \sigma T^4 \tag{3-3}$$

where T is the object's temperature expressed in Kelvin (K) and σ is called the Stefan-Boltzmann constant given as 5.67×10^{-11} kW/m^2-K^4. This formula applies to a perfect radiator, or **blackbody**.

blackbody
a radiator emitting the maximum possible radiation

emissivity
that property (0 to 1) that gives the fraction of energy emitted relative to a perfect radiator

Objects do not necessarily radiate at the maximum output given by Equation (3-3). Surface effects and absorbing effects reduce this output. The property that gives the fraction of real output to the maximum value is called **emissivity**, symbolized by the Greek letter ε. For solid and liquid surfaces, ε is typically 0.8 ± 0.2 for fire radiation applications. For gases or flames, ε depends on the thickness of the flame. Flames radiate due to discrete pockets of radiation from gases (fuels and combustion products) and a full spectrum of radiation frequencies from soot particles. The flame emissivity can be estimated from the formula,

$$\varepsilon = 1 - \exp(-\kappa l) \tag{3-4}$$

absorption coefficient
that property that pertains to the amount of radiation absorbed per unit length

where κ is the **absorption coefficient** and l is the flame thickness. The absorption coefficient is a property of the flame and is a measure of how easy it is for the radiation to penetrate the flame. For example, x-rays have a relatively low absorption coefficient for body tissue, implying easy penetration of the x-rays. For turbulent flames, κ typically ranges from 0.1 to 1.0 m^{-1}. For common fuel flames having a thickness of 2 m or more, their ε would be nearly 1, the maximum radiative output. But for very large fires, soot can actually obscure the flame which can reduce the radiant heat flux to the surroundings. This situation is contrary to the notion that large fires produce more radiant heat flux than smaller fires.

Let us consider a calculation for the incident radiant heat flux caused by a fire or a hot surface to a remote target object such as depicted in Figure 3-6. The source of the radiation can be considered to be a hot surface or a flame at temperature T_2. The radiant heat flux that is received at a target a distance, c, away will be reduced from that emitted by T_2. The fraction of energy reduced is called the **configuration factor**, designated by F_{12}. It represents the fraction of rays that can be seen from the target relative to the emitting object (see Figure 3-7). F_{12} depends on distance, c; the size of the source as object 2; and on the orientation of both object and target. For the arrangement depicted in Figure 3-6, F_{12} can be found from the chart given in Figure 3-8. Typical heat transfer references contain many charts for difference configurations.

configuration factor
fraction of radiation received by a target compared to the total emitted by the source

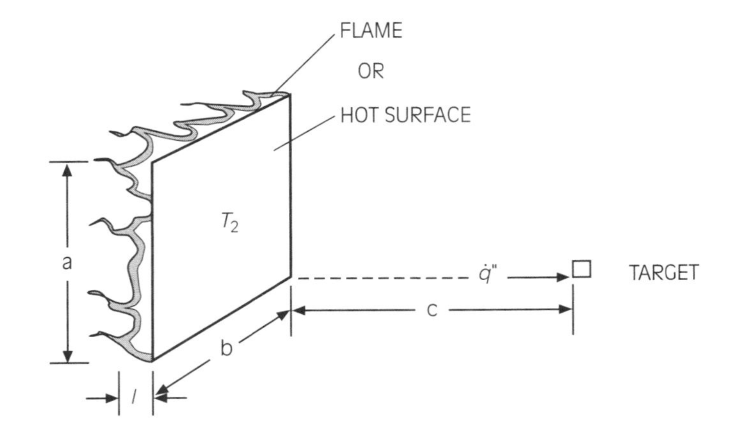

Figure 3-6 *Radiation to a target object from a flame or hot surface at temperature* T_2.

The heat flux received by the target object is given as

$$\dot{q}'' = \varepsilon\, \sigma\, T_2^4\, F_{12} \tag{3-5}$$

Let us consider an example based on Figure 3-6.

Example: Find the heat flux due to radiation from a wood fuel flame 1 m tall (a), 0.5 m wide (b), and 0.5 m thick (l) at a distance 3 m away (c). κ is 0.8 m^{-1} and the temperature is assumed to be 800°C. (It is very difficult to make precise radiation calculations because the fuel flame properties are approximate and the data are scarce. Nevertheless, we can make estimations.)

$$T_2 = 800°C + 273°C = 1073K$$

To determine F_{12}:

$$x = a/c = 1/3 = 0.333$$

$$y = b/c = 0.5/3 = 0.167$$

From Figure 3-8, $F_{12} = 0.017$.

$$\varepsilon = 1 - e^{-\kappa l} = 1 - \exp[-(0.8\ m^{-1})(0.5\ m)] = 0.33$$

$$\dot{q}'' = \varepsilon\, \sigma\, T_2^4 F_{12} = (0.33)(5.559 \times 10^{-11}\ kW/m^2K^4)(1073\ K)^4(0.017)$$
$$= 0.42\ kW/m^2$$

Compute this again for $c = 1$ m, to see the difference.

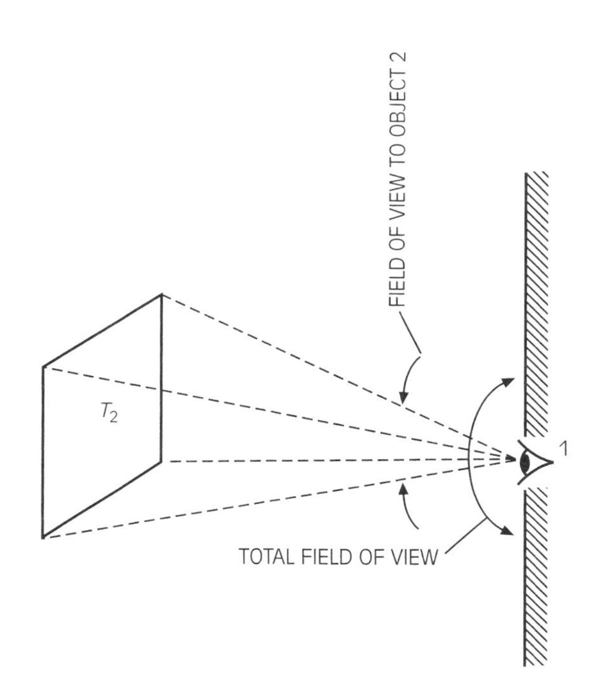

Figure 3-7 F_{12} *represents the fraction object 2 makes in the field of view from 1.*

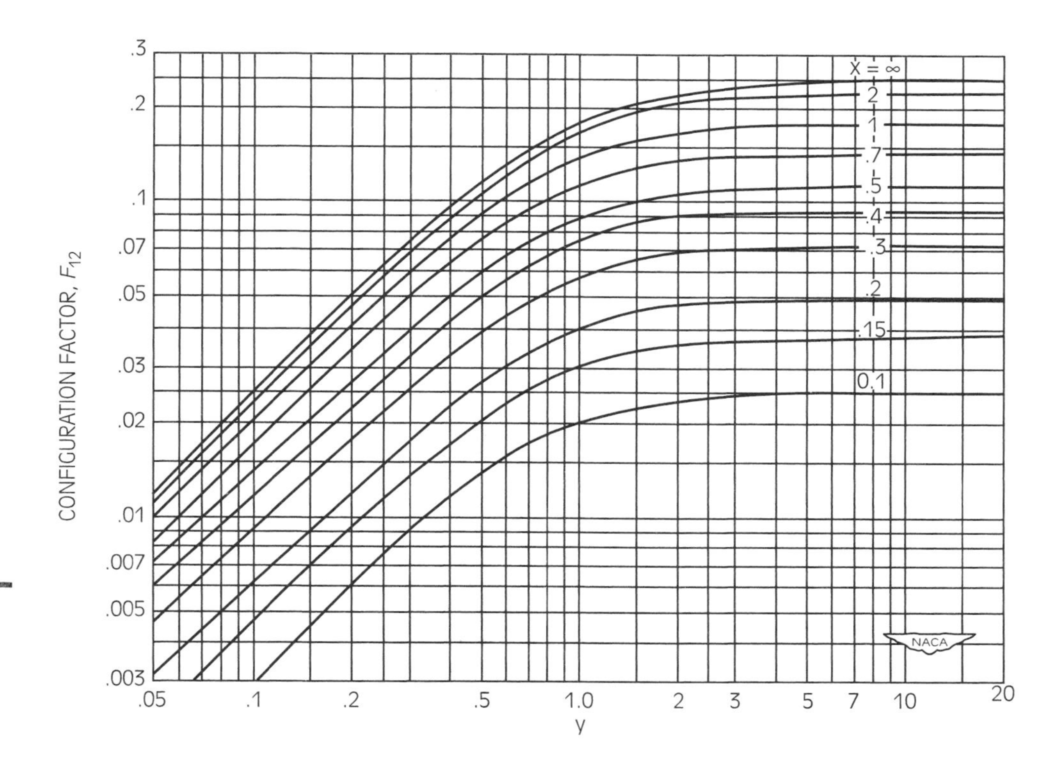

Figure 3-8 *Configuration factor,* F_{12}*. After Hamilton and Morgan, Ref. 2.*

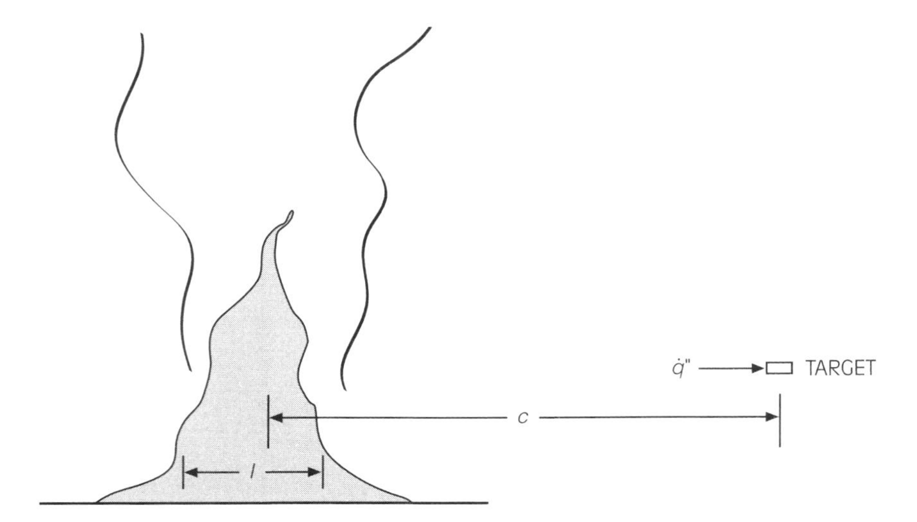

Figure 3-9 *Radiative heat flux to a target from a point-source such as flame.*

This computation of radiant heat flux to a target is very important to assess potential damage and the possibility of a remote ignition. Since the computation depends on flame shape and properties, it can be tedious if not impractical. Fortunately there is an alternative formula that is accurate as long as it is applied more than two flame diameters from the flame, i.e., $c > 2l$ (see Figure 3-9). The formula is

$$\dot{q}'' = \frac{X_r \dot{Q}}{4\pi c^2} \tag{3-6}$$

where $\dot{Q}$ = combustion energy release rate of fire (kW), and X_r is the fraction of energy radiated relative to the total energy released. X_r is not a constant for a given fuel, but generally varies from about 0.15 to 0.60 for low sooting fuels such as methane to high sooters, such as polystyrene. Tables 3-3 and 3-4 give some typical values. These are not respectively universal constants.

Equation (3-6) is based on uniform radiant emission from a "point-source" approximating a flame. All of the radiant energy released, $X_r\dot{Q}$, is uniformly received over a sphere of radius, c, from the flame. The area of the sphere is $4\pi c^2$ which completes the deduction of Equation (3-6). This simple result has been shown to have good accuracy as long as $c > 2l$.

For example, in the previous calculation for the wood flame we can make an alternative estimate. Select $X_r = 0.35$ (based on Figure 3-11). The 1-m high wood flame corresponds to about 25 kW (we will establish such a relationship in Chapter 7). Therefore, from Equation (3-6):

$$\dot{q}'' = \frac{(0.35)\,(25\text{ kW})}{4\pi(1\text{ m})^2} = 0.696\,\frac{\text{kW}}{\text{m}^2}$$

Table 3.3 *Typical radiative energy fractions,* X_r.

Fuel (l > 0.5 m)	X_r (%)
Methanol, methane	15–20
Butane, benzene, wood cribs	20–40
Hexane, gasoline, polystyrene	40–60

Source: Various.

Table 3-4 *Specific radiation fraction of combustion energy for hydrocarbon pool fires showing dependence on diameter.*

Hydrocarbon	Pool Size (m)	% Radiative Output/Combustion Output
Methanol	1.2	17.0
LNG on land	18.0	16.4
	0.4 to 3.05	15.0 to 34.0
	1.8 to 6.1	20.0 to 25.0
	20.0	36.0
LNG on water	8.5 to 15.0	12.0 to 31.0[a]
LPG on land	20.0	7.0
Butane	0.3 to 0.76	19.9 to 26.9
Gasoline	1.22 to 3.05	40.0 to 13.0[a]
	1.0 to 10.0	60.1 to 10.0[a]
Benzene	1.22	36.0 to 38.0
Hexane	—	40
Ethylene	—	38

[a]In these cases, the smaller diameter fires were associated with higher radiative outputs.

Source: From Mudan and Croce, Ref. 3.

This is very similar to the result of 0.42 kW/m^2 previously estimated. The differences are due to the approximate nature of both formulas and the accuracy of the data we are using to represent the wood flame. But the differences are within the accuracy we can expect from such a calculation. In general, fire calculations such as these can have an accuracy as poor a ±50%, but usually ±30% . Their value lies not in perfect accuracy, but in the plausibility of the fire scenario they support. For example, both of these results give heat fluxes comparable to sunlight. We do not expect them to be fire hazards.

HEAT FLUX AS AN INDICATION OF DAMAGE

Heat flux causes objects to get hot and possibly damaged or ignited. As a benchmark, the radiant heat flux from the sun at the Earth's surface is nearly 1 kW/m^2, at most. Thresholds (minimum values) of heat flux to cause damage under fire conditions are listed below:

Pain to bare skin:	1.0 kW/m^2
Burn to bare skin:	4 kW/m^2
Ignition of objects:	10 to 20 kW/m^2

These threshold values will cause the results indicated after a long exposure of many seconds or minutes. This is illustrated from the work of Stoll and Greene[4] showing the threshold and time to cause damage to bare skin (Figure 3-10).

The heat flux under fire conditions depends on many factors as we have seen, but some illustrations can provide general guidance in quantifying the effects of fire.

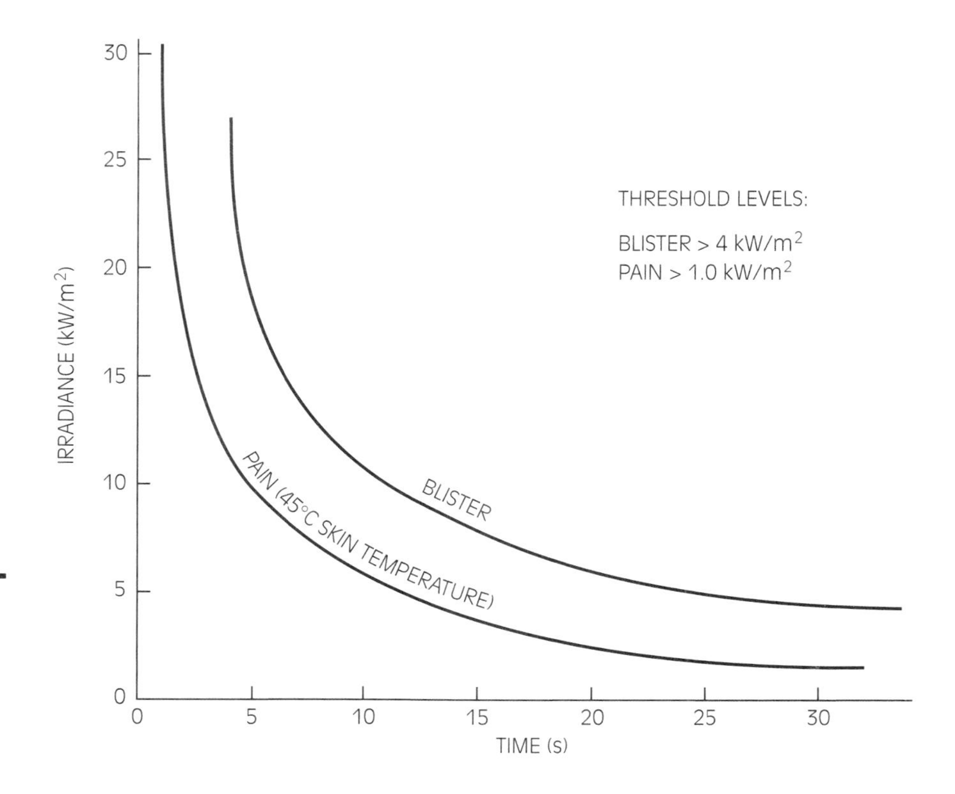

Figure 3-10 *Incident radiant heat flux (irradiance) effect on bare skin. After Stoll and Greene, Ref. 4.*

Figure 3-11 shows measurements of radiant heat flux received by a target approximately 1 m from wood or plastic cribs. The results are consistent with Equation (3-6) and yield a value of X_r between 0.30 and 0.40.

Incident floor heat fluxes due to these same wood and plastic cribs burning in a room with a doorway are shown in Figure 3-12. Despite the type of fuel and the variation of ventilation of air into the doorway, the results are almost only dependent on the average temperature of the smoke layer. This also applies to measurements of total (convective plus radiative) heat flux to a ceiling target as shown in Figure 3-13. As might be inferred by examining the field of view of both the ceiling and floor targets, the smoke layer dominates the field of view relative to the crib fire. This suggests that the smoke is the principal source of the heat flux as confirmed by its correlation with the smoke layer temperature. Moreover, the threshold of ignition (thin objects at 10 kW/m^2 and thick objects at 20 kW/m^2) suggests that objects begin to have a high propensity to ignite when the smoke layer temperature attains 400 to 600°C. This is why **flashover** is often associated with smoke layer temperatures at 500 to 600°C.

flashover
an event that can occur at a smoke temperature of 500 to 600°C in which flames suddenly fill a room

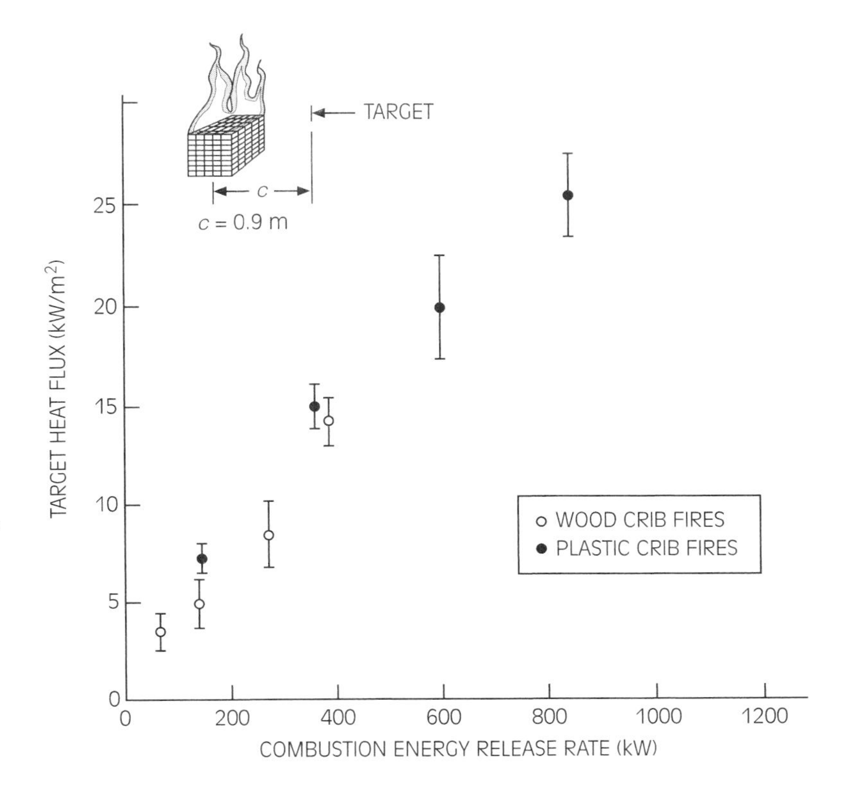

Figure 3-11 *Radiant heat flux to a target facing 0.9 m from the center of wood and plastic crib fires. After Quintiere and McCaffrey, Ref. 5.*

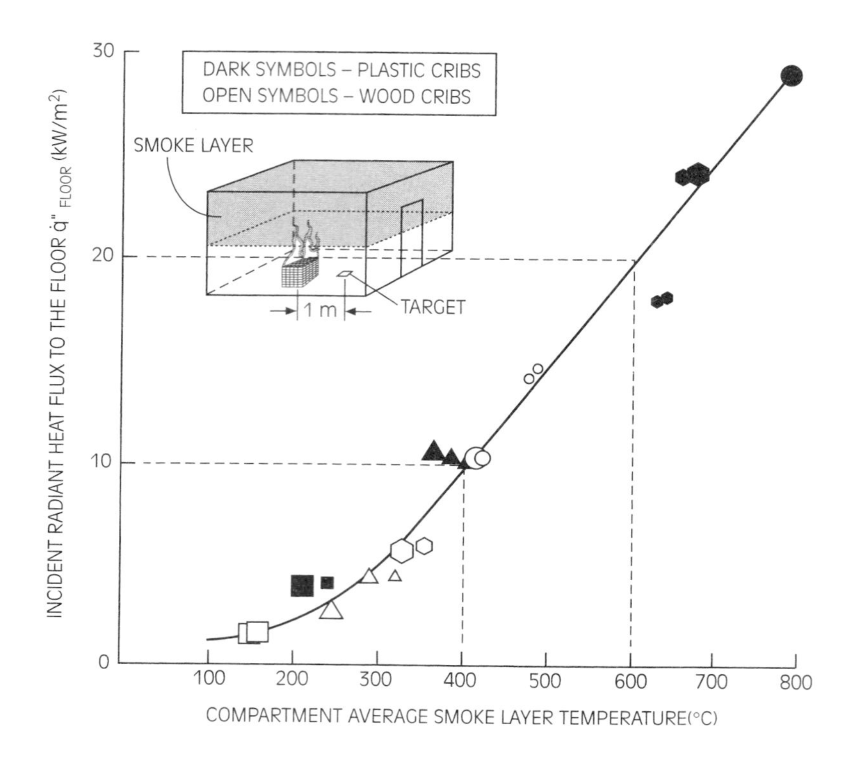

Figure 3-12 *Incident radiative heat flux to the floor in a compartment due to hot smoke. After Quintiere and McCaffrey, Ref. 5.*

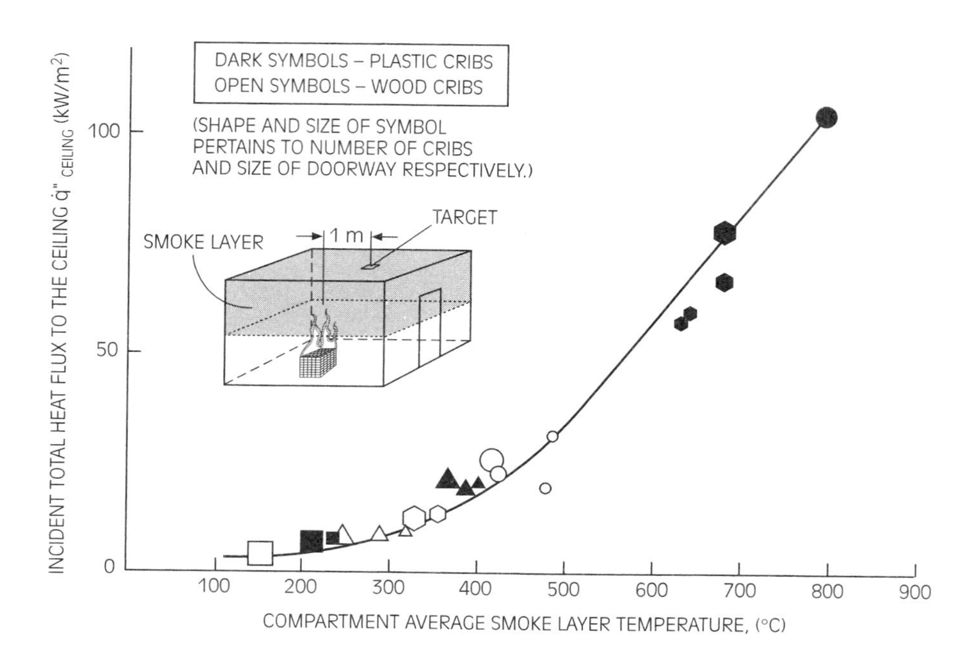

Figure 3-13 *Incident total ceiling heat flux in a compartment due to hot smoke. After Quintiere and McCaffrey, Ref. 5.*

Summary

Heat transfer is a significant process in fire. It accounts for the vaporization of the fuel, growth of the fire, and its damage.

There are three categories of heat transfer, (1) conduction, a molecular phenomena; (2) convection, conduction in a moving fluid; and (3) radiation, an electromagnetic phenomena. The laws of Fourier and Planck provide the bases for conduction and radiative heat transfer, respectively.

Heat flux, the flow rate of heat received per unit area by a target surface, is a key parameter in assessing the potential damage by a fire. Threshold heat flux levels are approximately 1 kW/m^2 for pain to bare skin; 4 kW/m^2 for a burn to bare skin; and 10–20 kW/m^2 for the ignitions of objects.

The latter normally corresponds to a smoke layer temperature in compartment fires of 500°C and is frequently taken as a benchmark of flashover initiation.

Review Questions

1. How long will it take heat to penetrate a 3-inch thick concrete wall?
2. Estimate the convective heat flux from a match flame to the wood. The flame is roughly 1900°C and the wood pyrolyzes at 350°C.
3. Estimate the convective heat flux from a gasoline turbulent pool fire to the evaporating gasoline. Gasoline evaporates at roughly 33°C during burning conditions. Its turbulent flame temperature is 800°C on average.
4. Compute the radiant heat flux 10 m from an 8 MW gasoline pool fire. Assume the fire is 3 m in diameter. If the fire diameter increases, do you expect the radiation emitted to go up or down?

True or False

1. Convective heat transfer is a type of conduction.
2. Conduction depends on molecules, radiation on electromagnetic energy, and convection on ether rays.
3. The emissivity of a surface should be less than one, but for a large flame it can easily be one.
4. Thermal conductivity and thermal diffusivity are properties of a solid controlling its heat transfer.
5. A threshold for a skin burn could be 4 kW/m^2, while at 10 kW/m^2 we could have ignition of clothing.

Activity

1. Find out how much time is needed to melt an ice cube. Look up the energy needed to melt ice (heat of fusion, J/g). Estimate the convective heat transfer coefficient for a cube sitting in still air. Note that the ice temperature is always 0°C, and record the room temperature. Compare your computed time to the actual time. Explain the differences.

References

1. D. Drysdale, *An Introduction to Fire Dynamics* (New York: Wiley, 1985), 36.
2. D. C. Hamilton and W. R. Morgan, *Radiant-Interchange Configuration Factors*, NACA Tech. Note 2836. (Washington, DC: December 1952), 78.
3. K. S. Mudan and P. A. Croce, "Fire Hazard Calculations for Large Open Hydrocaron Fires," chap. 3-11 in *SFPE Handbook of Fire Protection Engineering*, 2d ed., edited by P. J. DiNenno. (Quincy, MA: National Fire Protection Association, June 1995).
4. A. M. Stoll and L. C. Greene, "Relationship between Pain and Tissue Damage Due to Thermal Radiation," *Journal of Applied Physiology*, 14 (1959): 373–382.
5. J. G. Quintiere and B. J. McCaffrey, *The Burning of Wood and Plastic Cribs in an Enclosure: Vol. I*, NBSIR 80-2054 (Gaithersburg, MD: September 1980, National Bureau of Standards), 118.

Ignition

Learning Objectives

Upon completion of this chapter, you should be able to:

- Explain the difference between piloted and autoignition.
- Understand the concept of ignition temperature for solids.
- Use formulas to predict the ignition time of solids.

INTRODUCTION

In Chapters 4, 5, and 6 we discuss the essence of fire growth, which is composed of ignition, flame spread, and burning rate. These are distinct fire processes that may have some features in common, but must be put together to establish fire growth. We will clarify their common characteristics and their differences. This chapter discusses ignition, the start of fire growth.

PILOTED AND AUTOIGNITION

piloted ignition ignition of a flammable fuel–air mixture by a hot spot, spark, or small flame (pilot)

autoignition initiation of fire by chemical process inherent in the material; specific fuel concentration and temperature is usually needed

chemical kinetics refers to the rate of the chemical reaction

evaporation the process of gas molecules escaping from the surface of a liquid

Ignition was discussed in Chapter 2 for gaseous fuels. It has two forms: **piloted ignition** and **autoignition**. The former is the process of initiation and flame propagation in premixed fuel systems. The minimum condition for piloted ignition occurs at the lower flammable limit—that concentration of fuel that allows propagation with a small spark (pilot). We saw this process occur in Chapter 2 when the candle flame was blown out and the white vapor trail was then ignited. (What happens when we ignite the candle by applying a flame to the wick?)

The second form of ignition is autoignition, that occurs without any spark or flame source. The fuel must still be within a specific concentration range, and **chemical kinetic** processes must exceed the ability of the mixture to lose heat. Both piloted and autoignition occur in an identical fashion for the evaporated or decomposed fuel gases of liquid and solid fuels, respectively, as illustrated in Figure 4-1.

The liquid in Figure 4-1 evaporates at a sufficient rate to form a concentration near the pilot source at the lower flammable limit. The rate of **evaporation** is controlled by the liquid temperature. For example, water will evaporate relatively slowly under normal room temperatures. At that liquid temperature, or energy level, only a relatively few water molecules can break out of the surface and enter

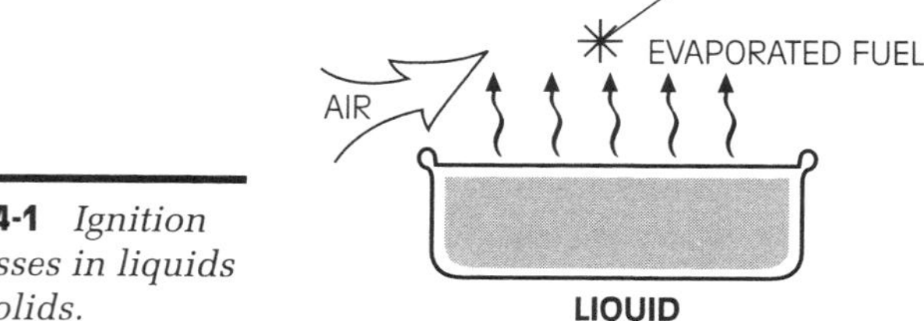

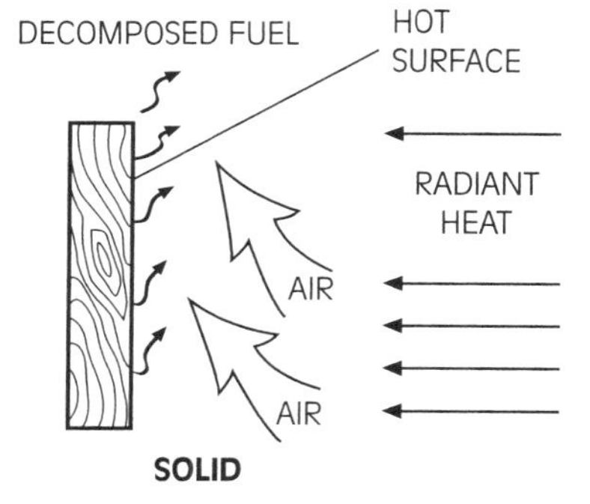

Figure 4-1 *Ignition processes in liquids and solids.*

humidity
the property of a water–air mixture that measures the amount of water present relative to the equilibrium concentration

flashpoint
the temperature of a liquid fuel, theoretically corresponding to the lower flammable limit of its evaporated vapor, and the point of piloted ignition

boiling point
the temperature at which a liquid can evaporate under normal atmospheric conditions

the adjacent air. The concentration of water molecules in the air is controlled by temperature and expressed by a measurement called **humidity**. At the surface of the evaporating liquid, the concentration of vapor is at equilibrium, and a maximum for that temperature. There the humidity is 100%, signifying no more evaporated vapor can be absorbed by the air. As the liquid temperature is increased, more molecules escape (evaporation) until 100°C (212°F) is reached when the humidity and the vapor concentration just at the surface are both 100%. At this point, no more increase in surface temperature or concentration can occur. This condition allows burning for liquid fuels. But for their ignition, the surface concentration needs only to achieve the lower flammable limit. At this concentration the corresponding surface temperature is called the **flashpoint** (T_{FP}) because the pilot initiates a flame propagation in the atmosphere above the liquid surface. In contrast, at the higher concentration of 100% vapor, the surface is at the **boiling point** (T_B). This distinction is shown in Table 4-1. Because normal atmospheric temperature is roughly 295 K (22°C), everything above ethanol in Table 4-1 can easily ignite under normal conditions. Kerosene and higher flashpoint liquid fuels need additional heating.

The process of autoignition is a different mechanism. For example, in Figure 4-1, the surface temperature of a solid such as wood must achieve the minimum temperature for spontaneous combustion of the adjacent gaseous fuel concentration. Note the autoignition temperatures of the liquid fuels are much higher than their flashpoints and boiling points. Indeed, all of the liquid would be vaporized at these temperatures. For that reason, we would expect (and find) similar results

Table 4-1 *Critical temperatures for liquid fuels.*

Liquid	Formula	T_{FP} (K)	T_B (K)	T_{AUTO} (K)[a]
Propane	C_3H_8	169	231	723
Gasoline	mixture	~ 228	~ 306	~ 644
Acrolein	C_3H_4O	247	326	508
Acetone	C_3H_6O	255	329	738
Methanol	C_3H_3OH	285	337	658
Ethanol	C_2H_5OH	286	351	636
Kerosene	~ $C_{14}H_{30}$	~ 322	~ 505	~ 533
m-Creosol	C_7H_8O	359	476	832
Formaldehyde	$C H_2O$	366	370	703

[a]Based on a stoichiometric mixture in a vessel.

■ NOTE

Autoignition temperatures are higher than piloted-ignition temperatures; both temperatures correspond to the liquid and solid exposed surface temperatures.

■ NOTE

Piloted ignition is probably the most common ignition process.

for solids. That is, autoignition temperatures are higher than piloted ignition temperatures, and both of these ignition temperatures correspond to the liquid and solid exposed surface temperatures.

Under actual fire conditions both autoignition and piloted ignition temperatures apply. However, the piloted ignition case is probably more predominant. Even the heating of items remote from the fire could cause a vapor trail to intercept the flames. This is a common scenario of arsonists creating delayed ignition using gasoline, or the cause of accidental fires due to spilling liquid fuels near a water heater pilot as illustrated in Figure 4-2.

Solids remotely heated by fire may char, causing a glowing ember at the surface. This ember can become the pilot. Moreover, embers made airborne by the main fire can become pilots. Consequently, the piloted ignition condition probably prevails most often.

Flashpoints for liquids can be measured with good precision and can also be computed from theory. But the (piloted) ignition temperature of solids is not so precisely determined. It depends on the decomposed fuel concentration and on the way the solid is heated. For example, under convective (hot air) heating, wood can ignite as low as 200°C, whereas under radiant heating the ignition temperatures range from 300° to 400°C. The formation of porous char may contribute to the differences. As a result, solid ignition temperatures are not likely to be published in the literature for each particular fire condition. Therefore, approximate temperatures must be used to make calculations based on available data. Often these data depend on the experimental conditions or the analysis used to derive them. Moreover, it is very difficult to directly measure the surface temperature up to the point of ignition. We shall now examine some predictive models for solid fuel ignition based on the use of effective or approximate **ignition temperatures** and other properties.

ignition temperature the surface temperature needed to cause ignition in solids

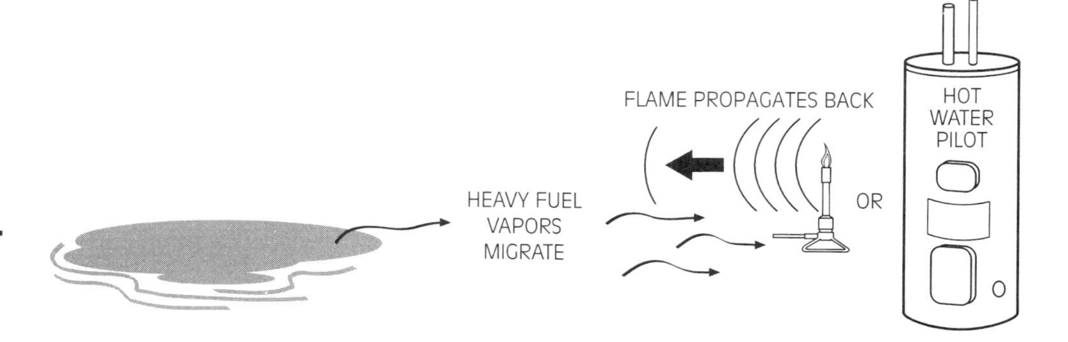

Figure 4-2 *Evaporating liquid; remote ignition.*

IGNITION TIME FOR SOLID FUELS

We can expect piloted ignition temperatures to range from approximately 250° to 450°C, with autoignition temperatures exceeding 500°C. Fire applications comprise a wide range of materials and commercial products; it is not possible to have exact data for each. However, test data can be very valuable if they allow us to obtain effective properties for items ranging from curtains to countertops. These effective properties enable us to estimate ignition times under a range of fire conditions.

■ NOTE
Surface temperature is the key parameter in determining ignition of solids (and liquids).

As we have seen, the key parameter in determining ignition of solids (and liquids) is the surface temperature. If and when that surface attains the ignition temperature is key to the ignition time. Not only will this depend on the way the heating occurs and the properties of the material, but on whether the material is "thin" or "thick." Thin is not just a physical dimension, but applies if the temperature is uniform throughout the thickness during the heating process. Except for very slow heating conditions (e.g., long-term heating of wood by steam pipes), "thin" is a physical thickness of 1 to 2 mm, or less than 1/16 inch. So, single sheets of paper, drapes, and garments could be representative of thin objects in a fire environment. Anything above this dimension, or even wallpaper on plasterboard is considered a thick assembly. The paper and plaster must be considered a composite assembly.

Ignition of Thin Objects

Two cases of thin heating are shown in Figure 4-3. An object of thickness l heated from one side and *perfectly* insulated on the other (a); or the same object of thickness $2l$ symmetrically heated and cooled (b). The incident heat flux is given

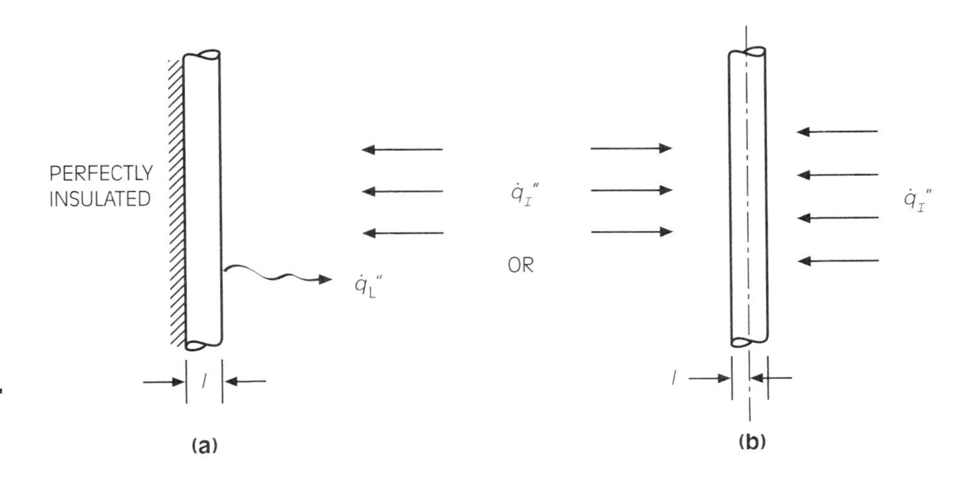

Figure 4-3 *Heated thin objects.*

by $\dot{q}''_I$, and it may be considered radiative, or from a flame. There is a heat loss flux $\dot{q}''_L$, which can be both radiative and convective. If an object is to increase in temperature, $\dot{q}''_I$ must be greater than $\dot{q}''_L$, and this difference should be great in order to achieve the ignition temperature. Let us represent this difference as $\dot{q}''$, a net heat flux.

This net heat flux is converted into internal energy of the solid material raising its temperature. How fast that happens depends on the capacity of the solid to store energy. This energy storage capacity is measured by ρcl, the product of ρ, the density, c, the **specific heat**, and l, the thickness.

specific heat
property that measures the ability of matter to store energy

As a result, the temperature of the thin solid will increase in time (t) as

$$T = T_\infty + \frac{\dot{q}''t}{\rho cl} \tag{4-1}$$

where T_∞ is the initial temperature. This formula only applies perfectly under short times and high heating rates. The temperature response over time is more completely shown in Figure 4-4 as well as the ignition temperature (T_{ig}). When T_{ig} is reached, we have ignition (piloted or auto, depending on our selection of the value for T_{ig}). The determination of the time to ignite is shown in the figure; but under low heating conditions, T_{ig} may never be achieved. Note, there is **critical heat flux** that just causes the attainment of the T_{ig} value. For the high heating rate cases above the critical heat flux, the ignition time can be computed using Equation (4-1) to find

critical heat flux
a threshold level below which ignition (or in another context, flame spread) is not possible

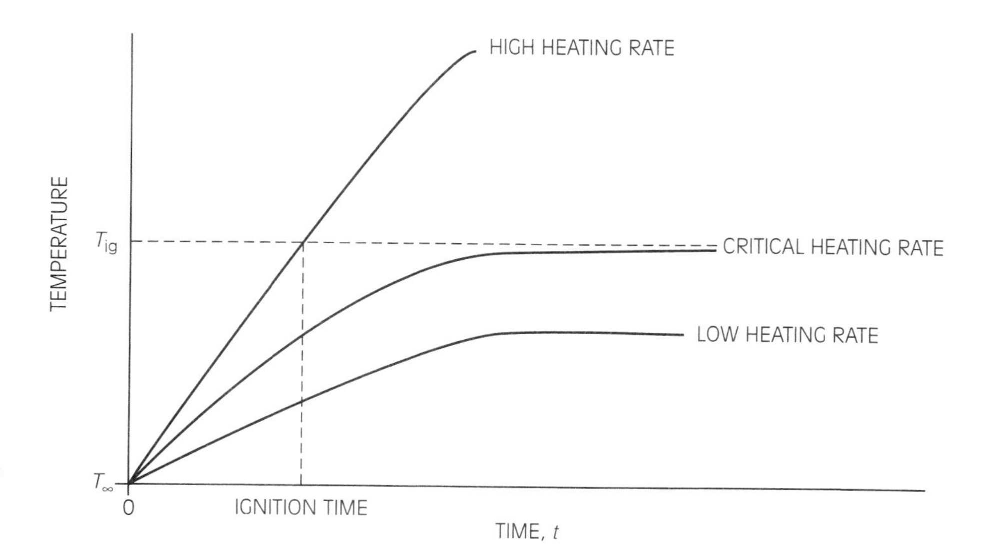

Figure 4-4
Temperature rise.

$$t_{ig} = \rho c l \frac{(T_{ig} - T_{\infty})}{\dot{q}''} \tag{4-2}$$

(for $\dot{q}''_I > \dot{q}''_{critical}$ and $\dot{q}'' = \dot{q}''_I - \dot{q}''_{critical}$).

Let us consider a typical material that might represent a drapery or a thin fabric on an insulating substrate, similar to a cushion or upholstered product. Representative properties for a cottonlike material are $\rho = 0.57$ g/cm^3, $c = 0.34$ cal/g-K, and $l = 1$ mm. For $T_{ig} = 300$°C, ignition times range from 5 to 25 s for a range of radiant heat fluxes between 40 and 10 kW/m^2, respectively. For $T_{ig} = 400$°C, these times only increase by about 25%. For this reason, these objects can ignite very quickly. Flashover conditions are sometimes designated with a heat flux to the floor of 20 kW/m^2 (indicative of room smoke layer temperature of 500°–600°C). So we see that most thin materials will rapidly ignite under this condition, consistent with our concept of flashover causing the fire to fill the room.

Ignition of Thick Materials

An equation similar to Equation (4-2) applies for thick materials (usually $l > 2$ mm). An approximate solution has been determined for thick (mathematically done for an infinitely thick solid) giving

$$t_{ig} = C(k\rho c)\left[\frac{(T_{ig} - T_{\infty})}{\dot{q}''}\right]^2 \tag{4-3}$$

where k is the thermal conductivity of the material and C is a constant, independent of material properties but somewhat dependent on heat flux[1,2]. For some approximate analyses, $C = \pi/4 = 0.785$ which holds for the ideal case of no surface heat loss and $C = 2/3 = (0.667)$, with heat losses. We shall use $\pi/4$. The variation in this constant is not so significant as the variations in $k\rho c$ and T_{ig}, as a result of approximate data due to issues of accuracy and precision. In general, k increases in solids as ρ (density) increases. Also k and c increase with temperature. Because the final temperature is T_{ig} (which might range from 200° to 500°C), k and c should vary considerably during the heating process. In Equation (4-3), constant values for $k\rho c$ are required at some appropriate average temperature. This value will be higher than that found for ordinary room temperature conditions.

Some typical ignition times for thick solids are listed in Table 4-2. We see that at 20 kW/m^2, a condition commonly used to imply flashover, it can take one or more minutes to achieve full involvement for common materials. But once higher room thermal conditions are achieved sufficient to ignite other materials, the transition time for full room fire involvement would be much shorter. During the flashover transition, the heat fluxes increase, with the heat fluxes in a room at

Table 4-2 *Typical ignition times of thick solids.*

Heat Flux (kW/m^2)	Time (s)	Material
10	300	Plexiglas, polyurethane foam, acrylate carpet
20	70	Wool carpet
	150	Paper on gypsum board
	250	Wood particleboard
30	5	Polyisocyanurate foam
	70	Wool/nylon carpet
	150	hardboard

full involvement attaining as much as 100 kW/m^2. As a result, the increasing heat fluxes reduce the times to ignition or flashover. Our simple formulas, Equations (4-2) and (4-3), do not take variable heat flux into account, but we certainly see that increased heat flux reduces the ignition time.

Typical ignition data are shown in Figure 4-5 for radiant heating of wood particleboard. The ignition times can be shown to agree well with Equation (4-3) ($t \sim (1/\dot{q}'')^2$), and at low heat fluxes there is no ignition. The minimum or critical radiant heat flux to cause ignition in a long time (greater than 5 minutes) is approximately 18 kW/m^2. Also shown in Figure 4-5 are measured ignition temperatures. These tough measurements were done with small thermocouples fixed to the wood surface. Moreover, flame flashing conditions at the surface make the onset of sustained ignition variable. Therefore, the scatter shown in the measured ignition temperatures can be expected. But a temperature of 350°C ± 50°C is easily accepted. If the same experiment is done with the flame heating as shown in Figure 4-6, the ignition times do not vary with flame energy (see Figure 4-7). The flame energy, as energy supply rate per width (kW/m) of wood heated, controls the flame height. The results show, despite variations in flame height that *this* flame causes ignition in 140 s ± 40 s. (Again, variations are due to experimental factors.) The concepts behind Equation (4-3) suggest that the flame heat flux under these conditions is approximately 22 ± 3 kW/m^2. This heat flux is typical of thin wall flames less than 1–2 m high *independent of the fuel*. A conveniently surprising result.

A collection of ignition data for *thick* materials is shown in Figures 4-8 through 4-11. Table 4-3 gives the corresponding properties of these materials along with others. In addition to T_{ig} and $k\rho c$ in Table 4-3, we list the critical heat flux, $\dot{q}''_{critical}$, below which ignition is not possible and Equation (4-3) no longer applies.

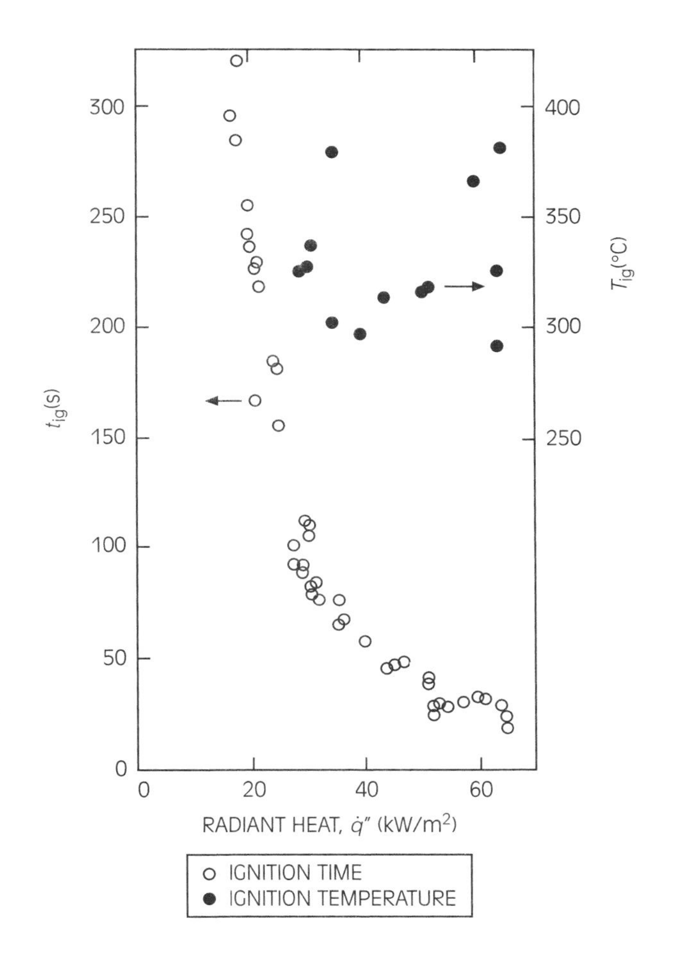

Figure 4-5 *Measurements of piloted ignition on particleboard by radiant heat flux. From Quintiere, Ref. 3.*

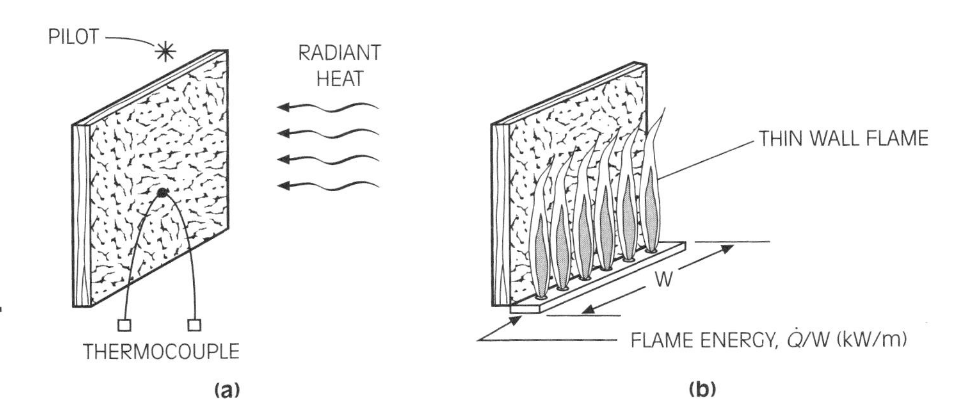

Figure 4-6 *Ignition experiments: a. radiation, b. flame.*

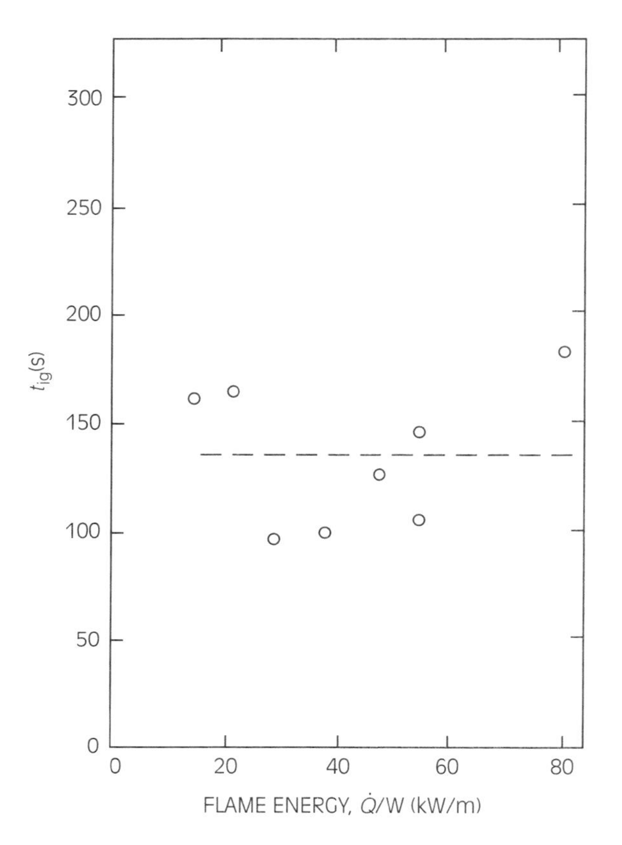

Figure 4-7 *Measurements of ignition on particleboard by a thin wall flame. From Quintiere, Ref. 3.*

Ignition Example

Determine the time for 1/2-inch-thick plywood to ignite subject to a flame heat flux of 25 kW/m^2. If a thin delaminated piece, insulated on the back, with a thickness of 0.5 mm is also heated, determine when it will ignite.

Estimated plywood properties:

$$T_{ig} = 350°\text{C}$$

$$k = 0.15 \times 10^{-3} \text{ kW/m-K}$$

$$\rho = 640 \text{ kg/m}^3$$

$$c = 2.9 \text{ kJ/kg-K}$$

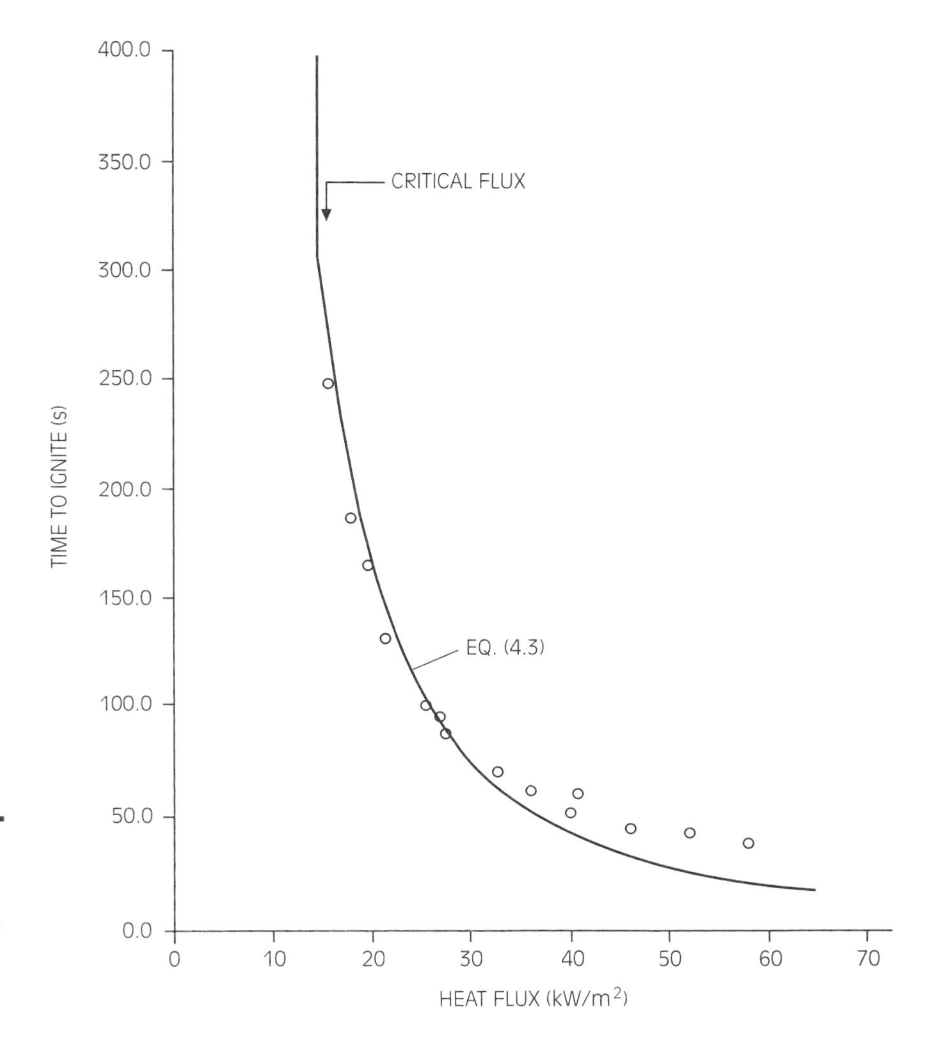

Figure 4-8 *Radiant ignition times of an asphalt shingle. From Quintiere and Harkleroad, Ref. 4.*

Thick case (1/2 in.)

$$
\begin{aligned}
t_{\mathrm{ig}} &= C(k\rho c)\left[\frac{(T_{\mathrm{ig}} - T_{\infty})}{\dot{q}''}\right]^2 \\
&= \frac{3.14}{4}(0.15\times10^{-3})(640)(2.9)\left[\frac{350-20}{25}\right]^2 \\
&= 38 \text{ s.}
\end{aligned}
$$

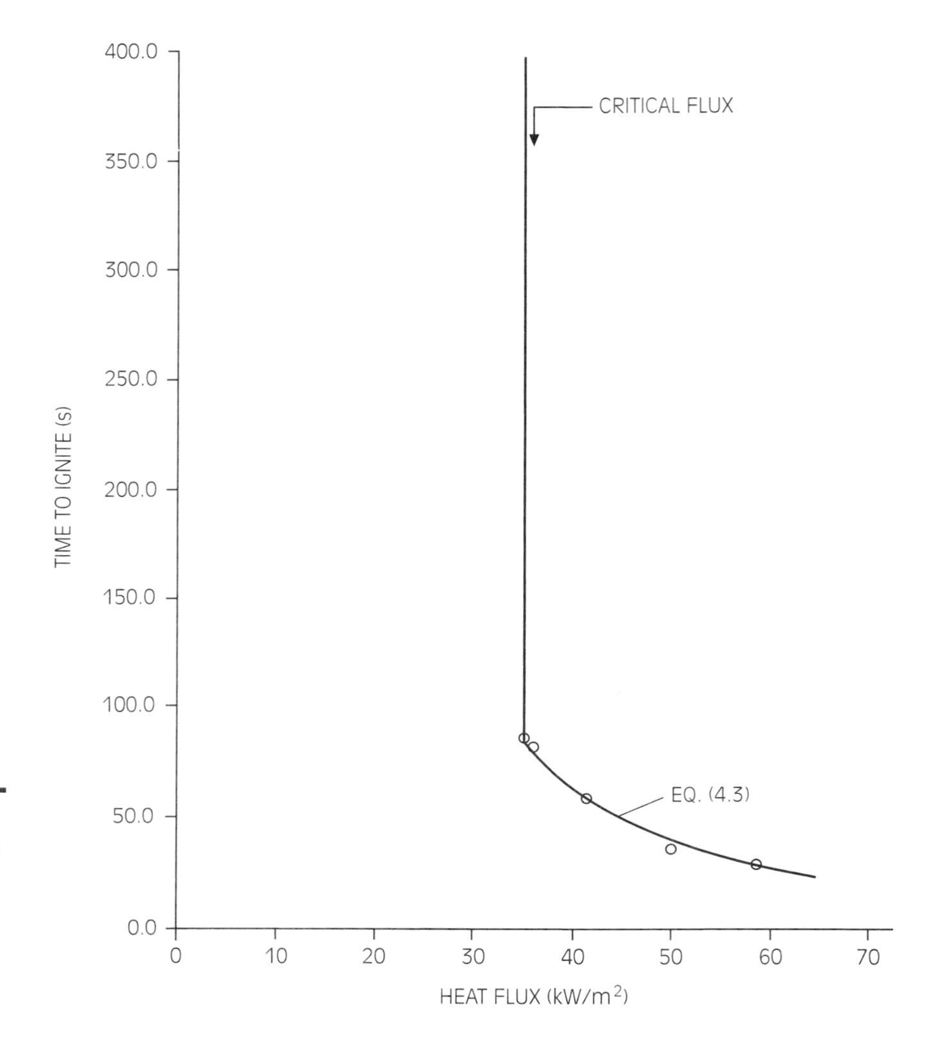

Figure 4-9 *Radiant ignition of paper on 1.27 mm gypsum board. From Quintiere and Harkleroad, Ref. 4.*

Thin case (0.5 mm)

$$t = \rho c l \frac{(T_{\text{ig}} - T_{\infty})}{\dot{q}''}$$

$$= (640)(2.9)(0.5 \text{ mm} \times 10^{-3} \text{m/mm}) \left[\frac{350 - 20}{25} \right]$$

$$= 12 \text{ s.}$$

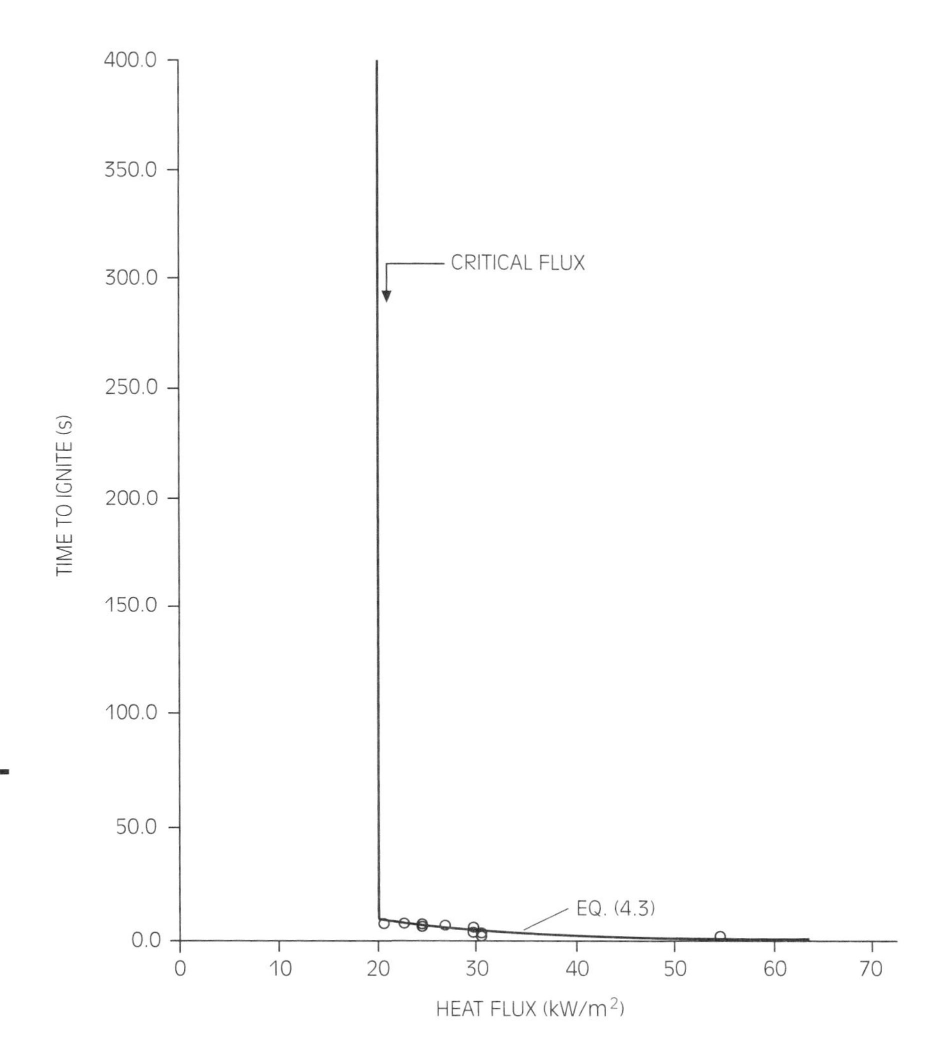

Figure 4-10 *Radiant ignition of 2.54 cm rigid foam plastic, polyurethane. From Quintiere and Harkleroad, Ref. 4.*

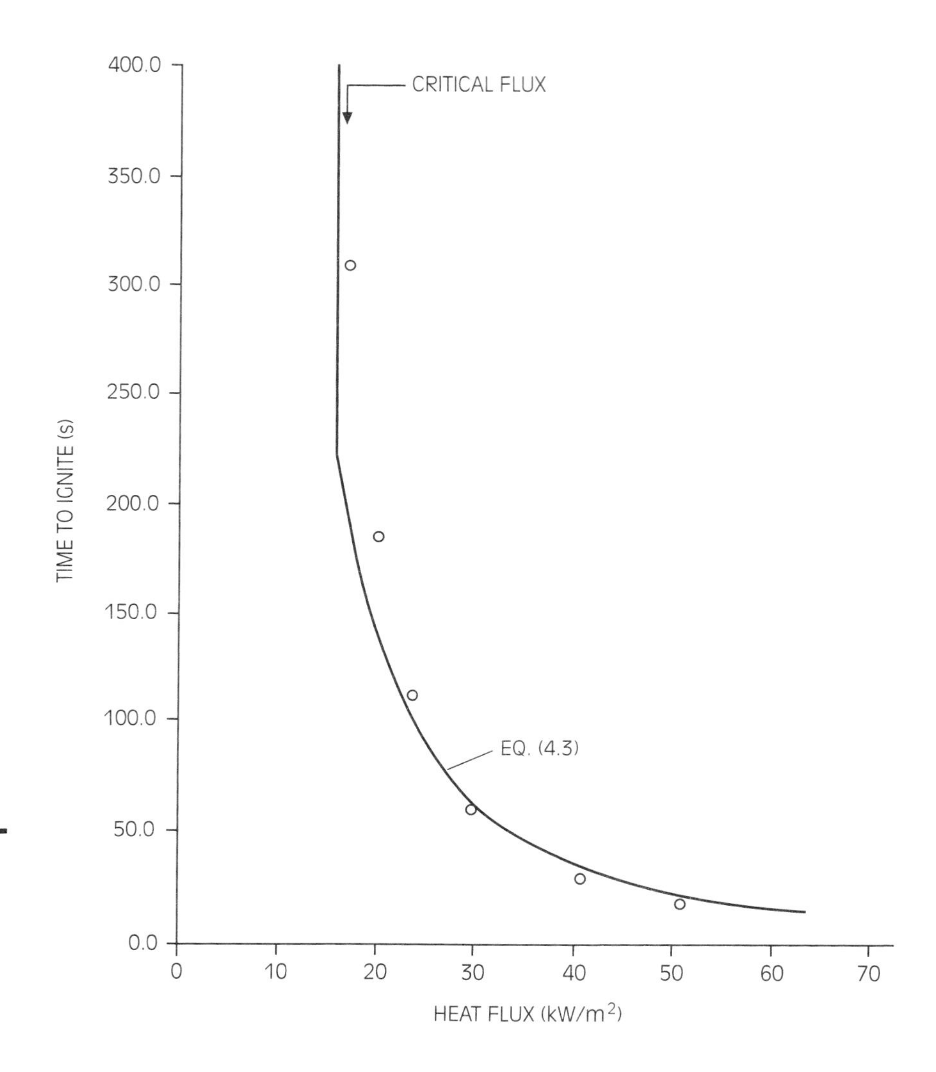

Figure 4-11 *Radiant ignition of 1.27-cm plywood. From Quintiere and Harkleroad, Ref. 4.*

Table 4-3 *Ignition properties.*

Material	$k\rho c$ $(kW/m^2K)^2s$	T_{ig} (°C)	$\dot{q}''_{critical}$ (kW/m^2)
Plywood, plain (0.635cm)	0.46	390.	16.
Plywood, plain (1.27cm)	0.54	390.	16.
Plywood, FR (1.27cm)	0.76	620.	44.
Hardboard (6.35mm)	1.87	298.	10.
Hardboard (3.175mm)	0.88	365.	14.
Hardboard (gloss paint), (3.4mm)	1.22	400.	17.
Hardboard (nitrocellulose paint)	0.79	400.	17.
Particleboard (1.27cm stock)	0.93	412.	18.
Douglas Fir particleboard (1.27cm)	0.94	382.	16.
Fiber insulation board	0.46	355.	14.
Polyisocyanurate (5.08cm)	0.020	445.	21.
Foam, rigid (2.54cm)	0.030	435.	20.
Foam, flexible (2.54cm)	0.32	390.	16.
Polystyrene (5.08cm)	0.38	630.	46.
Polycarbonate (1.52mm)	1.16	528.	30.
PMMA type G (1.27cm)	1.02	378.	15.
PMMA polycast (1.59mm)	0.73	278.	9.
Carpet #1 (wool, stock)	0.11	465.	23.
Carpet #2 (wool, untreated)	0.25	435.	20.
Carpet #2 (wool, treated)	0.24	455.	22.
Carpet (nylon/wool blend)	0.68	412.	18.
Carpet (acrylic)	0.42	300.	10.
Gypsum board, (common) (1.27mm)	0.45	565.	35.
Gypsum board, FR (1.27cm)	0.40	510.	28.
Gypsum board, wall paper	0.57	412.	18.
Asphalt shingle	0.70	378.	15.
Fiberglass shingle	0.50	445.	21.
Glass reinforced polyester (2.24mm)	0.32	390.	16.
Glass reinforced polyester (1.14mm)	0.72	400.	17.
Aircraft panel, epoxy fiberite	0.24	505.	28.

Source: From Quintiere and Harkleroad, Ref. 4.

Summary

Ignition for solid and liquid fuels begins with a mixture of the gasified fuel in air. At the correct concentration of fuel vapor, the mixture can propagate a flame at the lower flammable limit with a small energy source (pilot) or at a sufficient temperature alone (auto).

Two formulas were presented to predict the time to ignition of solids. These apply to a known heat flux incident on a thin (paper, fabric, etc.) or thick solid. The thin case involves the product $\rho c l$ while the thick case involves the product $k\rho c$. These comprise the thermal properties:

ρ = density

c = specific heat

k = thermal conductivity

and thickness, l.

For liquid fuels, ignition occurs at the minimum surface temperature known as the flashpoint. It corresponds to the lower flammable limit concentration of evaporated fuel at the surface.

Review Questions

1. A laminar match flame imparts roughly 60 kW/m^2 to a surface it contacts. How long would it take Douglas Fir particleboard (Table 4-3, page 79) to ignite under these conditions?
2. A small wastebasket fire in the corner against wood paneling imparts a heat flux of 40 kW/m^2 from the flame. The paneling is painted hardboard (Table 4-3, page 79). How long will it take to ignite the paneling?

True or False

1. Thin objects usually ignite more easily than thick objects.
2. Liquid fuels ignite at their boiling point.
3. Thick objects are more difficult to ignite as the material's density is increased.
4. Ignition temperature can be precisely measured for solid fuels.
5. Sunlight magnified 15 times could ignite thin paper.

Activities

1. Use an electric space heater to ignite pieces of newsprint. Find the distance from the heater where autoignition will just occur. By locating a lit match *above* the paper strip, record the time for piloted ignition to occur at the same location. Explain the results.
2. Experiment igniting small (very small, half dollar size) dishes of liquid fuels. Use ethenol or methanol and note its ease of ignition with a match flame. Cool the alcohol in a bath of ice water and see if it still ignites. Try igniting kerosene in the same way. Explain the results.
3. Try igniting a thick piece of wood (2 × 4) with a match flame. Does it ignite? Or, does it not spread a flame? Observe carefully. Are the corners or edges easier to ignite? Explain the results.

References

1. P. H. Thomas, "The Growth of Fire—Ignition to Full Involvement," chap. 5 in *Combustion Fundamentals of Fire*, edited by G. Cox, (London: Academic Press, 1995), 281.
2. A. Atreya, and M. Abu-Zaid, "Effect of Environmental Variables on Piloted Ignition," in *Fire Safety Science, Proceedings of the Third International Symposium*, edited by G. Cox and B. Langford, (London: Elsevier Applied Science, 1991), 183.
3. J. G. Quintiere, "The Application of Flame Spread Theory to Predict Material Performance," *Journal of Research of the National Bureau of Standards*, 93, no. 1 (Jan.–Feb. 1988): 61–70.
4. J. G. Quintiere, and M. Harkleroad, *New Concepts for Measuring Flame Spread Properties*, NBSIR 84-2943 (Gaithersburg, MD: National Bureau of Standards, November 1984).

Flame Spread

Learning Objectives

Upon completion of this chapter, you should be able to:

- Describe the different types of fire spread.
- Explain flame spread theory in terms of distance heated and ignition time.
- Given data, compute flame spread speeds on solids for different cases.
- Develop quantitative understanding of fire spread rates for many conditions.

flame spread
the process of advancing the fire front in air, along surfaces, or through porous solids

fire spread
the process of advancing a combustion front: smoldering or flaming

■ **NOTE**
Gravitational and wind effects are important to flame spread, and to fire growth in general.

wind-aided
refers to air flow in the same direction as the fire spread

opposed-flow
refers to air flow in the direction opposite to the fire spread

pyrolysis
the process of breaking up a substance into other molecules as a result of heating; also known as thermal decomposition

burning rate
the consumption of fuel mass per unit time

■ **NOTE**
Flame spread rate or velocity plays an important role in assessing the hazard in fire.

INTRODUCTION

Following ignition, that takes place in the fuel–gas air mixture, the next step in fire growth is surface flame spread. Of course, the propagation of flame through the gas mixture is a form of flame spread, but we are principally concerned with surface spread. Such spread involves the participation of the liquid or solid fuel. Its vaporization region must extend from the point of ignition. This spread does not always occur; so the label *non-self-propagating* has been applied to materials after testing. However, just because they are nonpropagating in a particular test does not mean they will not propagate under other fire initiation conditions. Such labels have been deceptive and often prove harmful to victims of fire accidents.

This chapter explains how a flame can spread over fuel surfaces or through a porous fuel matrix such as forest brush. Also, fire can propagate through solids as a smoldering as well as a flaming process. Although theory and formulas are presented, these are difficult to use in practical applications. However, we shall try to impart a level of fire spread speed to the various types of spread to give a quantitative perspective to the subject.

DEFINITIONS

Flame spread is that process in which the perimeter of the fire grows. It could include the process of remote ignitions if those ignition processes are continual. Examples of flame spread for a solid or liquid fuel can be seen in many fires. Specifically, we mean the extension of the burning region. It is not the flame extend, but the region volatilizing and supplying the fuel. In general, we can embrace the concept of **fire spread**, which is more general. Fire spread applies to the growth of the combustion process including surface flame spread, smoldering growth, and the fire ball in premixed flame propagation.

In flame spread, and in fire growth generally, gravitational and wind effects are important. The flows resulting from the fire's buoyancy or the natural wind of the atmosphere can assist (**wind-aided**) or oppose (**opposed-flow**) flame spread. Figure 5-1 depicts these modes. The flame spread velocity is defined as the rate of motion for the perimeter position x_p. The x_p denotes the extent of the **pyrolysis** or vaporization region.

Behind the spreading pyrolysis perimeter there is likely to be another perimeter behind which flaming or combustion has ceased. The region between these two fronts defines the principal flaming or pyrolysis region. The combustion of this pyrolysis area and the rate of pyrolysis is related to the primary hazard variable in fire. This variable, which we define as the **burning rate**, is directly connected to establishing the temperature, visibility, toxicity, and corrosivity of the fire. For this reason, flame spread rate or velocity play an important role in assessing the hazard in fire.

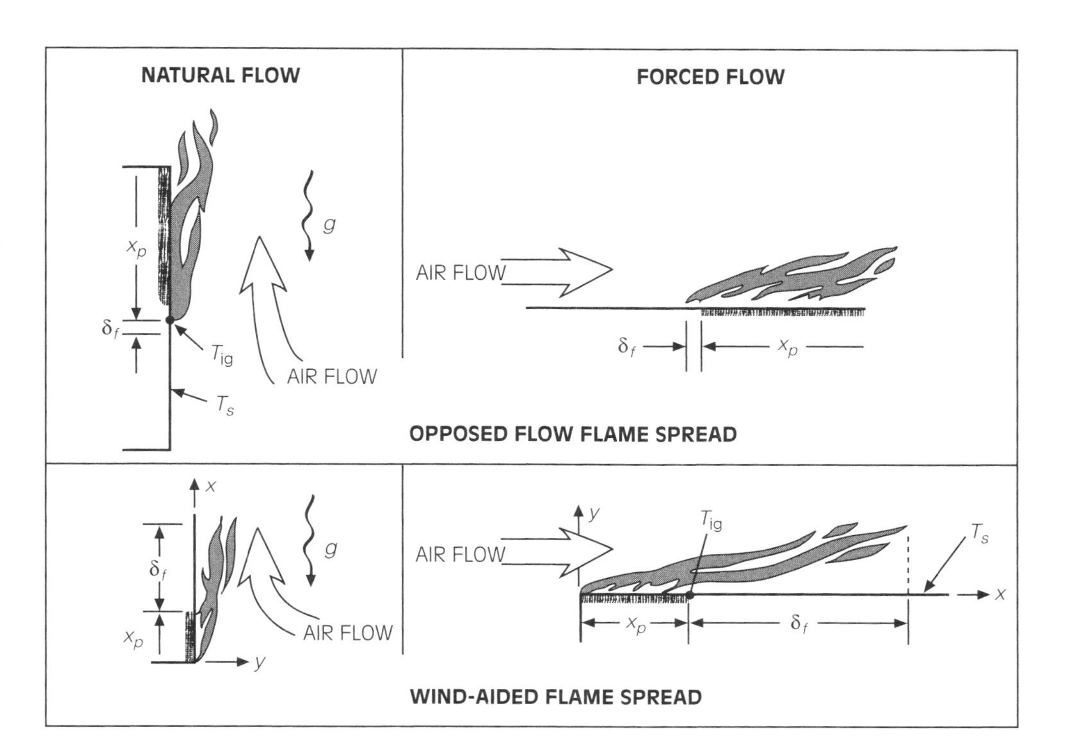

Figure 5-1 *Flame spread modes. Natural: Flow is induced solely by the buoyancy of the flow. Forced: Flow is caused by the meteorological wind or a powered fan. Opposed-flow refers to flame spread opposite to the air flow. Wind-aided refers to flame spread in the same direction as the air flow.*

THEORY

The concept of an ignition temperature is key to explaining flame spread in simple, but physically correct, terms. For flame spread, the pilot ignition temperature would apply because a pilot (the flame itself) is always present. If we examine Figure 5-1 again, we have labeled the position x_p at the ignition temperature, T_{ig}. The temperature ahead of the flame, not affected by direct heating from the flame, is labeled T_s. The distance along the surface affected by the flame's heat transfer is labeled δ_f. In general, the flames can heat the region ahead in many ways. These depend on the mode of spread—orientation, wind—and on the nature of the solid or liquid fuel. Symbolically, this process is illustrated in Figure 5-2. The illustrations depict steady flame spread. An observer *riding* on the flame front, at position x_p, sees the new fuel coming toward it at the flame spread velocity, V, or the velocity the flame has in spreading into the fuel, which is at rest. The flame spread

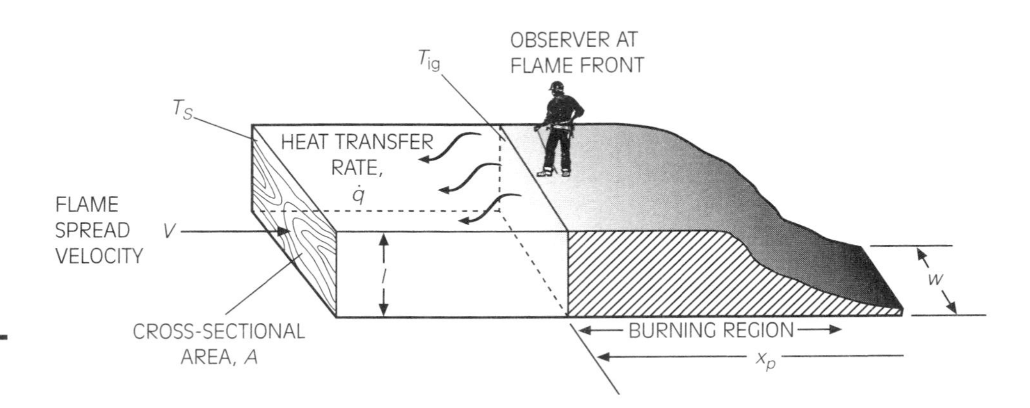

Figure 5-2 *Flame spread model.*

velocity formula states that the rate of energy supplied to this newly heated fuel, bringing it to temperature T_{ig}, is equal to the net heat transfer rate from the burning region, $\dot{q}$. Hence, we have a balance for spread:

[Rate of energy required] = [Rate of heat supplied]

$$\rho VAc\,(T_{ig} - T_s) = \dot{q} \tag{5-1}$$

where ρ = the fuel density,
c = the fuel specific heat,
A = the cross-sectional area, wl and,
T_s = the fuel temperature beyond the range of the flame's heat.

The flame spread velocity is then:

$$V = \frac{\dot{q}}{\rho cA(T_{ig} - T_s)} \tag{5-2}$$

Most specific cases of flame spread can be derived from this formula by more carefully describing $\dot{q}$ and A. We shall consider some approximate formulas; these are in agreement with more complex analyses beyond the scope of this text. These formulas should be considered as methods to make estimates, but not with high accuracy under all conditions. Again, orientation, wind, and the nature of the fuel all make a difference.

SPREAD ON SOLID SURFACES

For spread over solid surfaces, it has been shown that the most significant heat transfer rate is at the surface over a length δ_f as shown in Figure 5-1. Therefore,

$$\dot{q} = \dot{q}'' \, \delta_f \, w \qquad \textbf{(5-3)}$$

where $\dot{q}''$ is the flame forward heat flux and w is the width of the fuel as shown in Figure 5-2. The flow area, A is wl, where l is the fuel's thickness. If we substitute these formulas into Equation (5-2), we obtain

$$V = \frac{\dot{q}''\delta_f}{\rho c l(T_{ig} - T_s)} \qquad \textbf{(5-4)}$$

natural flow refers to air flow induced by buoyancy

forced flow refers to air flow produced by wind or a fan

Again, referring to Figure 5-1, V is the rate of movement of position x_p. Orientation and **natural flow** or **forced flow** conditions affect the heating distance δ_f and the associated flame heat flux $\dot{q}''$. Measurements have shown that for downward spread on a wall, the heat flux is approximately 70 kW/m^2 over a δ_f of approximately 1 mm. For upward flame spread on a wall, the heat flux is generally 25 kW/m^2 over the luminous flame region for flames extending from 0.20 to 2 m. These latter results appear to be independent of the wall fuel; so, whether we have a gasoline saturated wood wall, or the wood wall itself, the incipient heat flux from the flame to the wall is about 25 kW/m^2.

Just as for the ignition of solids, we must distinguish between thick and thin solids. The surface heat transfer from the flames can only penetrate to a certain depth. This heating in depth takes place during the time to heat the surface from its original temperature T_s to the ignition temperature T_{ig}. This heating time is, in fact, the ignition time caused by the heat flux of the spreading flame.

For a thin solid ($l < 2$ mm), the ignition time, t_{ig}, is given by Equation (4-2) which allows us to write Equation (5-4) as

$$V = \delta_f / t_{ig} \qquad \textbf{(5-5)}$$

since the thin solid ignition time is

$$t_{ig} = \frac{\rho c l(T_{ig} - T_s)}{\dot{q}''}$$

For a thick solid, the entire thickness l is not heated during the ignition time. Recall, that ignition time is given by

$$t_{ig} = \frac{\pi}{4} k\rho c \left[\frac{T_{ig} - T_s}{\dot{q}''}\right]^2$$

the thick solid ignition time, as given by Equation (4-3). From our discussion on heat conduction in Chapter 3, the actual depth heated in time can be estimated as $4\sqrt{\alpha t_{ig}}$, where α is the thermal diffusivity, $k/\rho c$. The spread velocity for the thick solid is also given by Equation (5-5) with this thick solid ignition time used. Most flame spread problems can be put into the form of Equation (5-5). The solution then depends on determining the heating length δ_f and the ignition t_{ig} caused by the flame heat transfer.

Composite materials, such as plasterboard with a paper face, should be regarded as a single *thick* material in determining ignition and spread. Although the paper on the plasterboard may be physically thin ($l < 2$ mm), the combination of the paper and plaster act as a *thick* material, because flame heat transfer from the paper burning will be conducted into the plaster. However, if the paper delaminates, then it will ignite and spread as a thin material.

Downward or Lateral Wall Spread

For downward or lateral flame spread on a wall surface, the combination of the δ_f and $\dot{q}''$, i.e., $(\dot{q}'')^2\,\delta_f$, for thick solids, depends on the air flow and on the fuel. For such spread in *still* air, this combination has been found to depend only on the fuel, and can be considered a fuel flame spread property. This property is designated by ϕ in the following formula:

$$V = \frac{\phi}{k\rho c(T_{ig} - T_s)^2} \tag{5.6}$$

ASTM E 1321
a standard providing a test to properly determine data for ignition and spread

This type of opposed-flow flame spread will only occur if the surface temperature is above a critical value. Such data can be obtained from **ASTM E 1321**: Standard Test Method for Determining Ignition and Flame Spread Properties.[1] Figure 5-3 shows a typical example of this type of surface flame spread. Below the minimum temperature for spread, 120°C, no spread occurs. Above this temperature, spread occurs at increasing velocity as the surface temperature is increased. This occurs in the ASTM E 1321 test by heating from an inclined external radiant panel burner. In actual fire conditions, this occurs by heat transfer from the hot smoke contained in a room. The flame heats a very small region (~1 mm) ahead of the lateral or downward spreading front. The smoke, in contrast, heats the wall to temperatures (T_s) that depend on the energy release rate of the fire. As this surface temperature increases, the surface flame can heat the wall to the ignition temperature (390°C) more quickly. Suppose a fire occurs in a room with the plywood wall shown in Figure 5-3. At some point the fire has sufficient heat flux (above the critical heat flux for ignition) to ignite the plywood. In order for lateral or downward flame spread to occur, the room fire conditions must heat the plywood to 120°C, its minimum temperature for spread. Here the spread rate is only 0.33 mm/s and achieves 3 mm/s, or about 7 in./min. at $T_s = 300$°C. This is how room fire conditions can accelerate fire spread over all burning objects. Equation (5-6) suggests that when the surface temperature of the plywood reaches 390°C, its ignition temperature, the speed would be infinitely fast. This assumption is not correct because as the plywood reaches its ignition temperature, the gaseous pyrolyzed fuel at the surface would be within its flammable limits. Consequently, the flame would

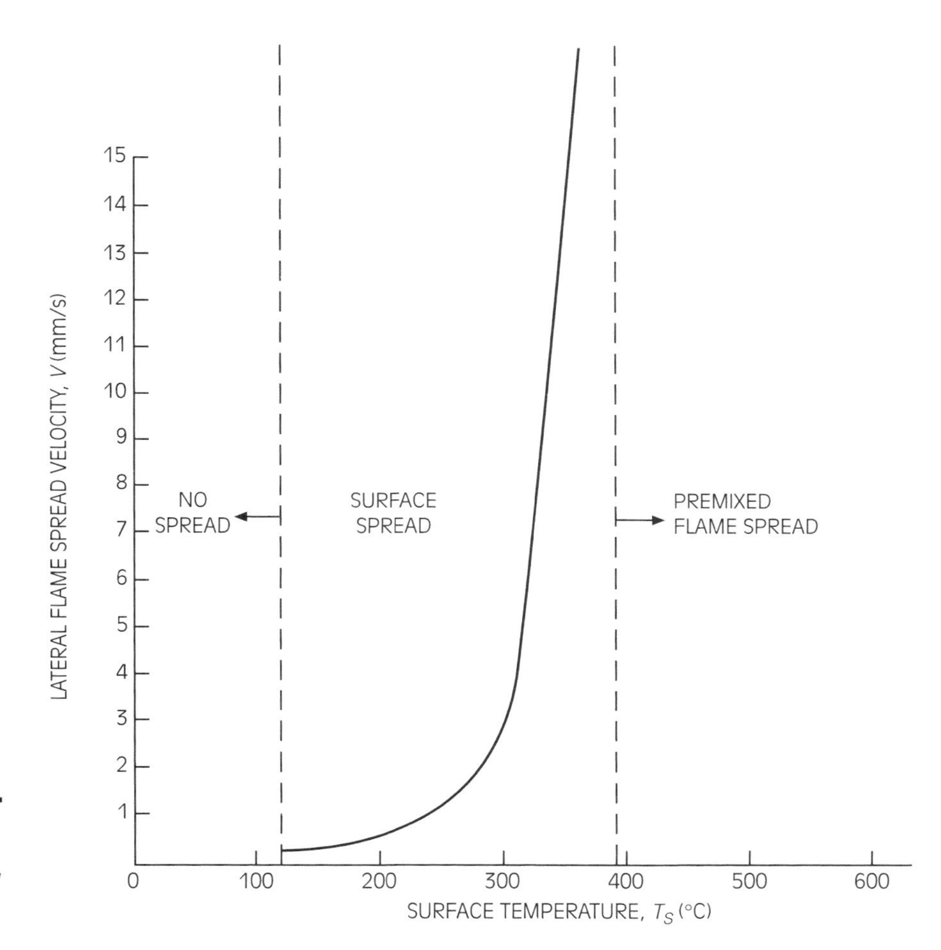

Figure 5-3 *Lateral flame spread based on a test of plywood in still air.*

propagate through the gas mixture near the surface with the appropriate premixed flame speed (probably 30–100 mm/s).

Typical data taken from ASTM E 1321 for lateral flame spread in still air is shown in Table 5-1. These data can be used to construct plots as depicted in Figure 5-3. The data should not necessarily be generalized to similar materials because their compositions may not be exactly the same, or fire retardants may have been included. However, they are generally representative of common construction or interior finish materials. The T_{ig} and $k\rho c$ values would apply to other computations in ignition and flame spread, but the ϕ and $T_{s,\,min}$ apply only to the lateral spread conditions in still air.

Table 5-1 *Lateral flame spread data from ASTM E 1321.*

Material	T_{ig} (°C)	$k\rho c$ $\left(\frac{kW}{m^2K}\right)^2 s$	ϕ $\left(\frac{kW^2}{m^3}\right)$	$T_{s,\,min}$ (°C)
Wood fiber board	355	0.46	2.3	210.
Wood hardboard	365	0.88	11.0	40.
Plywood	390	0.54	13.0	120.
PMMA	380	1.0	14.4	<90.
Flexible foam plastic	390	0.32	11.7	120.
Rigid foam plastic	435	0.03	4.1	215.
Acrylic carpet	300	0.42	9.9	165.
Wallpaper on plasterboard	412	0.57	0.8	240.
Asphalt shingle	378	0.70	5.4	140
Glass reinforced plastic	390	0.32	10.0	80

Source: From Quintiere and Harkleroad, Ref. 2.

Upward or Wind-Aided Spread

Flame spread in the direction of the air flow is generally referred to as wind-aided spread. Upward flame spread on a wall or flame spread under a ceiling in still air is also termed *wind-aided*. In these cases the *wind* is solely due to the buoyant flow caused by the fire itself. The heating needed to cause flame spread is mainly caused by the heat transfer from the extended luminous flame adjacent to the surface. The flame extends the distance δ_f as shown in Figure 5-1. This distance depends on the extent of burning. We will see in Chapter 7 that flame length depends principally on the energy release rate of the fire driving it.

Let us consider the scenario for upward flame spread on a vertical wall. An igniting fire, placed adjacent to a wall, "insults" it with a sufficient heat flux to ignite the wall over a heated length x_p. At that point the combined fuel from the igniting fire and the wall section ignited produce a flame height x_f. The flame extension δ_f is $x_f - x_p$. Our flame spread velocity at the onset of spread is given by Equation (5-5), where the ignition time (t_{ig}) depends on the flame heat flux at that time. As the wall fire continues to spread, it will be less and less dependent on the igniting fire. Indeed, the igniting fire may burn out. Also, the wall may burn out as well. Both of these extinctions will reduce the height of the flame, x_f. Alternatively, the wall material may contain sufficient fuel (e.g., plywood compared to wallpaper on plasterboard) to cause x_f to continue to increase. If x_f depends directly on x_p, the flame speed will accelerate, because x_f out runs x_p. For this reason, upward, or wind-aided flame spread in general, depends on the nature of the igniting fire and the combustion properties of the wall. Of course, if this upward flame spread scenario occurs in a room, the smoke layer will increase T_s causing the surface velocity to also increase as in downward or lateral spread. But unlike downward spread where δ_f is small and nearly constant (~1 mm), for upward spread δ_f can sharply increase or decrease due to combustion and flow properties that affect flame length, x_f. As a result, flame spread up a wall, under a ceiling, or driven by air flow in a ventilation duct are all different. Our ability to write perfect formulas for each case is not yet possible, so we are left with only the framework of Equation (5-5) at this time. Unlike the case of opposed-flow spread where δ_f is constant, these wind-aided cases will be unsteady, because δ_f will change as the flame propagates. Spread will accelerate or stop. Figure 5-4 gives examples of upward spread on corner wall materials subject to a 100-kW ignition source with an igniting heat flux of 40 kW/m^2.[3] Typical upward spread rates can range from 1 to 200 cm/s.

NOTE
Flame spread up a wall or under a ceiling, or driven by air flow in a ventilation duct will all be different.

SPREAD THROUGH POROUS SOLID ARRAYS

Flame spread can take place through porous solid fuels such as forest brush and debris. Even wind-aided blown flame spread through arrays of urban dwellings can be considered in this class of flame spread. For this class of spread the forward

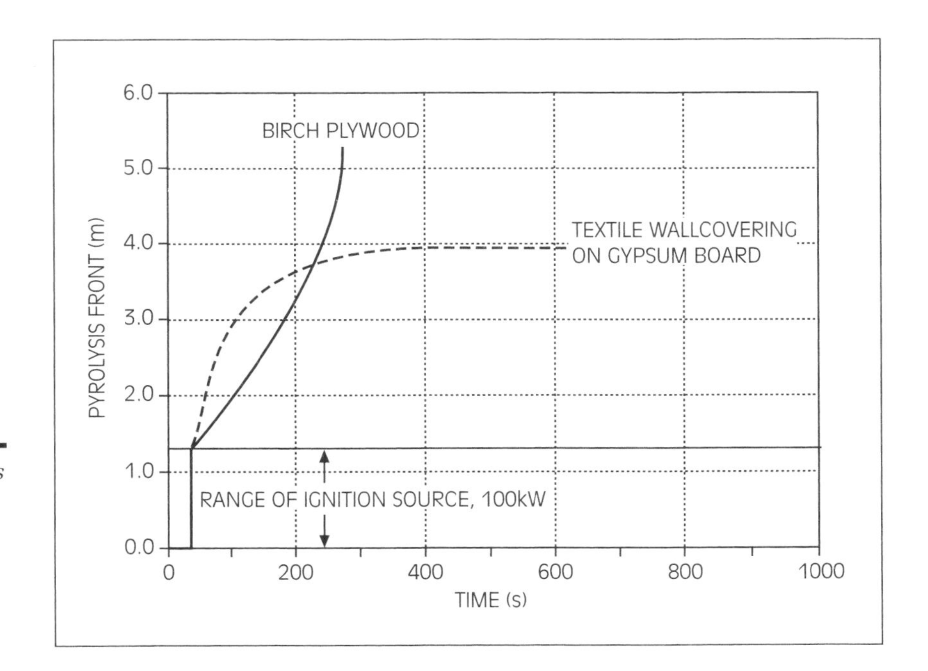

Figure 5-4 *Examples of computed upward flame spread on a wall due to a 100 kW line source. After Quinteire, Ref. 3.*

heat flux to cause spread is within the array, not outer surface spread. Consider Figure 5-5 in which the heat flux is aligned with the spread velocity acting over area *wl*. Let the fuel mass in the array divided by the volume of the array be defined as the bulk density, ρ_b. If the fuel elements are thin so that they heat uniformly, then since $A = wl$ from Equation (5-2) we obtain for the array:

$$V = \frac{\dot{q}''}{\rho_b c(T_{ig} - T_s)} \tag{5-7}$$

The heat flux and the thermal properties for forest materials are nearly constant and V inversely depends on ρ_b. This has been found to hold true for woody elements over a wide range of thicknesses. Thomas[4] reports that:

$$V = C/\rho_b \tag{5-8}$$

Figure 5-5 *Flame spread through a porous array.*

where V is in m/s,
ρ_b is in kg/m^3, and
C is a constant of roughly 0.07 kg/m^3 for wildland fuels and 0.05 kg/m^3 for wood cribs of sticks to 3 cm in diameter.

For wind-aided flame spread through porous arrays, Thomas[5] reports Equation (5-8) becomes

$$V = (1 + V_\infty)C/\rho_b \tag{5-9}$$

where V_∞ is the wind spread in m/s.

Figures 5-6 and 5-7 show the accuracy of these simple formulas for different porous wood-based fuel arrays. Equation (5-9) is a good approximation up to a

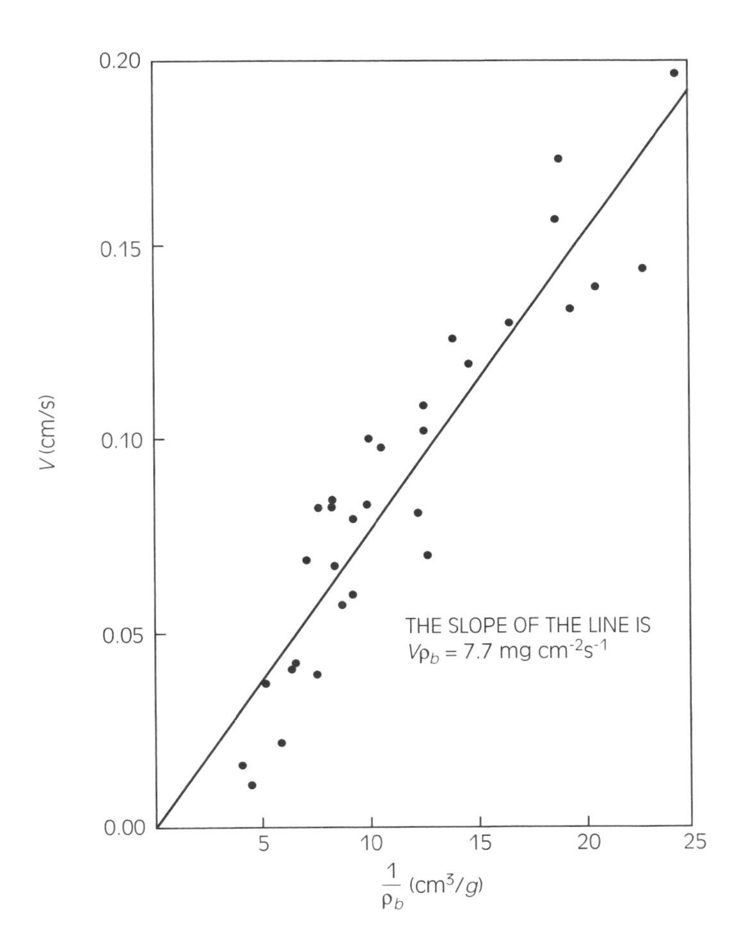

Figure 5-6 *The effect of bulk density on fire spread. After Thomas, Ref. 4.*

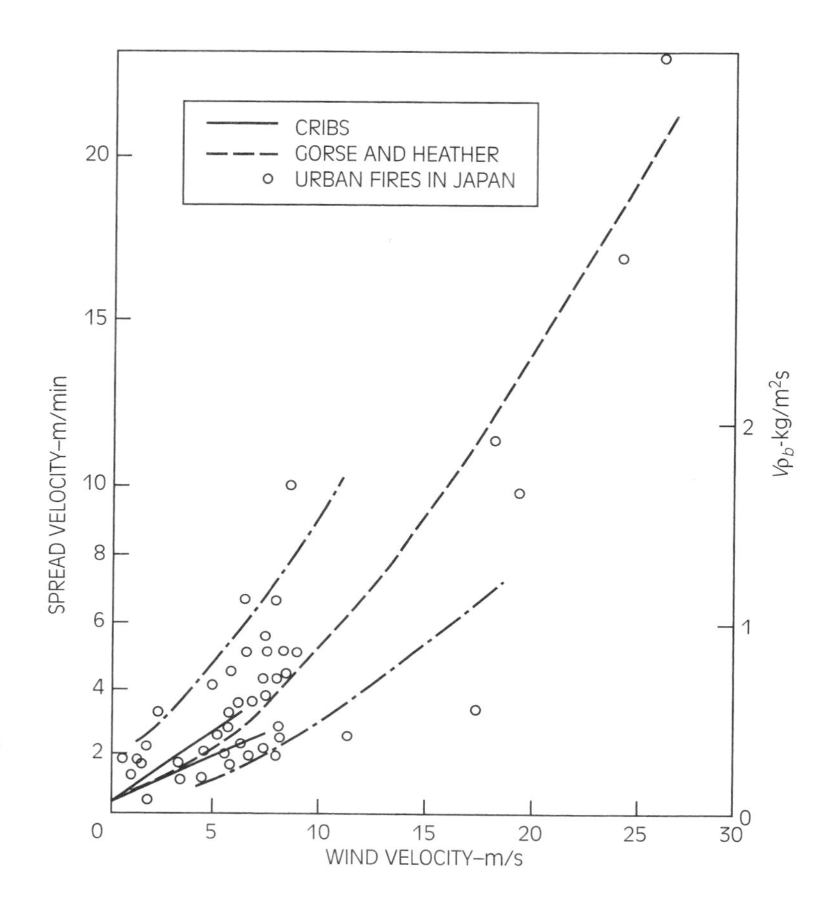

Figure 5-7
Comparison of rates of fire spread. After Thomas, Ref. 5.

wind speed of 4 m/s. Urban fire data for conflagrations of Japanese cities is also shown in Figure 5-7 where the bulk density of the dwellings has been taken as 10 kg/m^3 in order to plot the data. Figure 5-8 shows more recent results of fire spread through part of Kobe following the Hyogo-ken Nambu earthquake of 1995.

SPREAD ON LIQUIDS

Surface flame spread on liquids is similar to the mechanisms discussed for surface flame on solids. However, the liquid differs in that motion can be induced in the liquid due to the spreading flame. The fact that the surface of the liquid is hotter

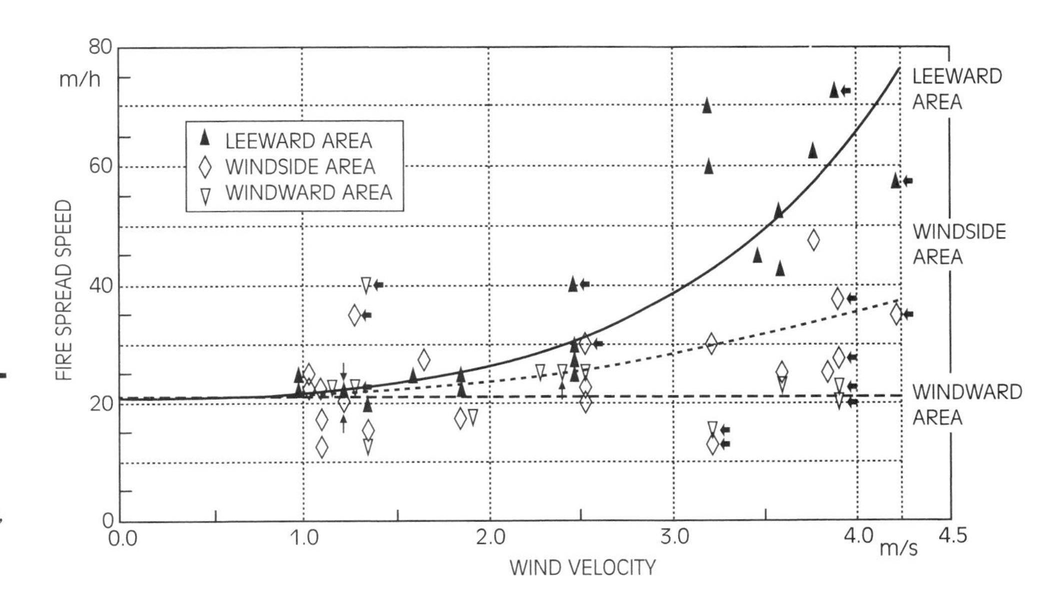

Figure 5-8 *Wind velocity affecting fire spread speed following the earthquake in Kobe, Japan. From Nagano, Ref. 6.*

viscous
refers to fluid friction

surface tension
a force within the surface of a liquid

under the flame than the cooler liquid upstream leads to a buoyant flow that can influence the flame spread process. Also, motion can be induced in the liquid by **viscous** or **surface tension** effects. Surface tension variation is the chief mechanism that makes liquid surface flame spread different from solids. Because surface tension decreases with temperature, the cooler liquid, ahead of the vaporization position x_p at the ignition temperature T_{ig}, has a higher surface tension. The heat conducted along the liquid surface between T_{ig} and T_s over distance δ_f results in a surface tension that *pulls* flame into the cooler region. Here, we are talking about flame spread on a horizontal liquid pool unaided by wind. This surface tension effect would not be a significant mechanism in wind-aided spread on liquid pools because surface flame heat flux would dominate. For flame spread on a liquid pool in still air, the surface tension effect causes significant heat transfer through the liquid ahead of the flame. This liquid heat transfer augments the surface heat transfer occurring from the air interface. Therefore, from Equation (5-2), the liquid flame spread speed is augmented over what we would expect for a solid comparable in properties. This increase is due to these buoyant and surface tension flows in the liquid fuel.

Figure 5-9 shows experimental results of Akita[7] for a 2.6 cm by 1.0 cm deep pool of methanol. The ignition temperature for methanol is the same as the flashpoint, which is 11°C. At that temperature, a fuel air mixture exists at the surface that can be ignited with a *piloted* flame or small spark. The flame speed results are plotted against the original liquid temperature T_s, similar to Figure 5-3 for ply-

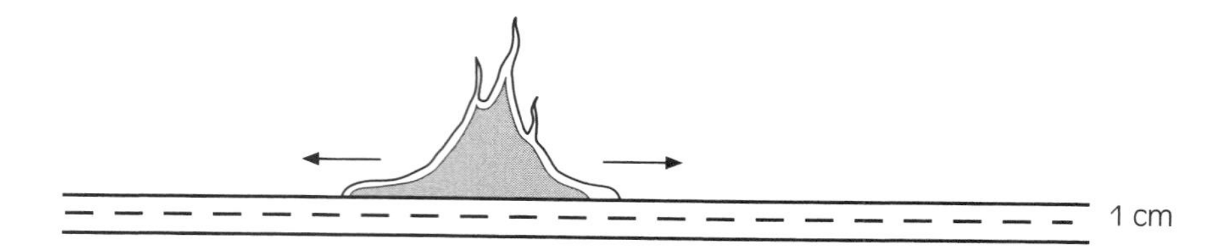

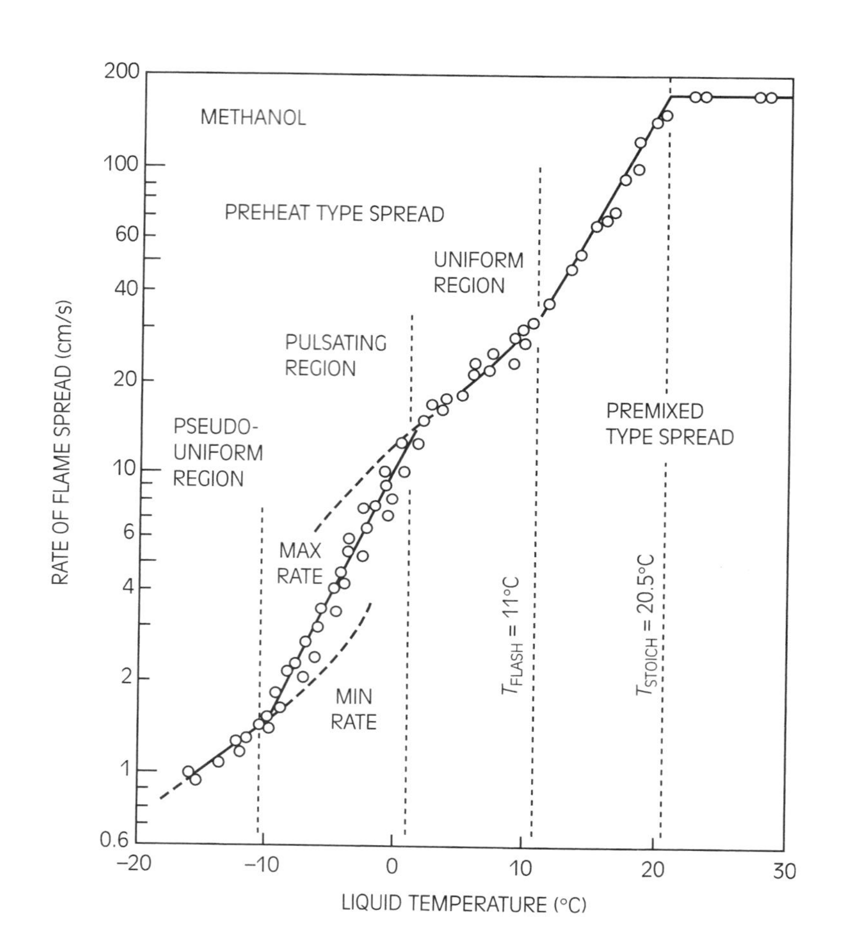

Figure 5-9 *Rate of spread on a liquid pool as function of its temperature. After Akita, Ref. 7.*

wood. The results are distinctly different because of a pulsating region that applies to liquid spread. Here the spread rate can oscillate between the maximum and minimum values shown. Also, in Figure 5-9 the premixed gas phase velocities are explicitly shown for temperatures above the flashpoint. Such results were not shown for the plywood in Figure 5-3.

TYPICAL FIRE SPREAD RATES

■ NOTE
Flame spread is a complicated phenomena of fire growth that does not easily lend itself to quantitative analysis.

As we have seen, flame spread is a complicated phenomenon of fire growth that does not easily lend itself to quantitative analysis. Formulas do exist, but they are usually lacking in information for their variables and for fuel properties. To put fire spread rates in perspective, Table 5-2 lists typical values that may be useful in characterizing different spread phenomena. Even without the application of formulas, the table data can usefully characterize fire spread phenomena.

STANDARD TEST METHODS

ASTM E 84
a standard providing a test (Steiner tunnel test) developed by Underwriters Laboratories to examine wind-aided flame spread under ceiling mounted materials

A number of flame spread-based test methods exist for materials, some of which form the basis for fire safety regulations in the United States. Most of these tests produce relative indexes for assessing spread and do not support basic data for the computation of flame spread. Some can even be misleading because the test performance may not reflect realistic fire scenarios. Other tests have played a traditional role in regulating fire safety and have been supported on an empirical bed of data (see Table 5-3). The **Steiner tunnel test (ASTM E 84)** is a wind-aided flame spread test with the material on the ceiling of a duct. It gives a flame spread rating that is normalized at a rating of 100 for the rate associated with burning red

Table 5.2 *Typical flame spread rates.*

Spread	Rate (cm/s)
Smoldering	0.001 to 0.01
Lateral or downward spread on thick solids	0.1
Wind driven spread through forest debris or brush	1 to 30
Upward spread on thick solids	1.0 to 100.
Horizontal spread on liquids	1.0 to 100.
Premixed flames	10. to 100. (laminar) $\approx 10^5$ (detonations)

Table 5.3 *Empirical standard flame spread tests.*

ASTM Designation: E 84	Standard Test Method for Surface Burning Characteristics of Building Materials
ASTM Designation: E 162	Standard Test Method for Surface Flammability of Materials Using a Radiant Heat Energy Source
ASTM Designation: E 648	Standard Test Method for Critical Radiant Flux of Floor-Covering Systems Using a Radiant Heat Energy Source
ASTM Designation: E 1321	Standard Test Method for Determining Material Ignition and Flame Spread Properties

ASTM E 162
a standard providing a radiant fuel test to measure the flammability of materials under downward spread

ASTM E 648
a standard providing a flame test for floor-oriented materials

oak. Although **ASTM E 162** is a downward flame spread test, its measurements are processed to give similar index results to ASTM E 84. **ASTM E 648** is one of the few tests that actually give results in engineering terms. It gives the minimum heat flux required for horizontal flame spread on a floor material. In contrast, ASTM E 1321, is not used for regulatory purposes, but generates engineering data for flame spread computations. Such data have been listed in Table 5-1. Suffice it to say that no standard test has been invented that directly gives a full set of engineering data needed to predict flame spread. This deficiency is a serious problem in assessing the fire growth hazard of new materials and their applications.

Summary

Fire spread can occur through gaseous fuel at incipient speeds of 10 cm/s reaching detonations in confined regions of 10^5 cm/s. On the other hand, surface flame spread can range from 1 to 100 cm/s on liquids and solids, especially if it is wind driven. Downward or lateral spread can have much lower speeds, and smoldering almost stands still at 0.001 to 0.01 cm/s. All of these speeds contribute to the fire growth and the extent of the burning region. The extent of the burning region is directly related to hazard and damage due to fire.

Theories have been described to illustrate the factors important in fire spread and to provide a framework for computations. Fire spread depends on many factors: the fuel, its orientation, the wind, the direction of spread, and other factors. Fire spread velocity can be described as the ratio of heated distance to ignition time. The heated distance depends on the *reach* of the fire's direct heat flux. It has been shown how such a simple theory can be applied to the many situations encountered in fire spread.

Review Questions

1. Data for a cardboard match is as follows:
   $k = 0.20$ W/mK
   $\rho = 550$ kg/m^3
   $c = 2.5$ J/g °C
   $\phi = 12$ kW^2/m^3
   $T_{ig} = 350$ °C
   Assume the flame imparts a heat flux of 50 kW/m^2. At normal room temperature, T_∞ = 20°C, compute the flame spread speed.
   a. For downward spread, the heating distance (δ_f) is 1.5 mm.
   b. For upward spread when the flame extends from the pyrolysis front by 2 cm.
2. A rigid foam plastic wall lining is ignited. The flame heat flux is 25 kW/m^2, and after several seconds extends 1.5 m from the ignition region. Compute the upward spread speed using data in Table 5-1.

True or False

1. Thickness is not a factor in flame spread.
2. As materials are heated by hot smoke in a fire, they will spread faster than at normal temperatures.
3. Flame spread on liquid fuels is the same as that on horizontal solids.
4. The bulk density or porosity of forest brush is an important variable in flame spread.
5. Winds do not change the spread rate, but can blow out the fire.

Activities

1. Tape the edges of strips of paper so as to suspend them from a wire clothes hanger. The tape is used for mounting and holding the paper in place. Ignite the vertical strip to observe upward and downward spread. See if you can clock the time for the pyrolysis front to spread over the strip. Moisten the paper with alcohol (ethyl or methyl) and repeat the tests. Explain your results.
2. Place two 2 × 4's about 1 foot high together. Find the gap thickness that will allow flame to spread up or down the gap by using a match flame size ignitor. Explain your results.
3. Obtain a shallow metal or glass tray about 6 to 10 in. long and 1 in. wide. Separately pour kerosene, then methyl or ethyl alcohol into the tray about 1/4 in. in depth. Ignite one end and record the spread time. *Do not* let the alcohol sit long. Explain the results.

References

1. "Standard Test Method for Determining Material Ignition and Flame Spread Properties," ASTM E 1321, in *1996 Annual Book of ASTM Standards, Sec. 4, Construction*, (West Conshohocken, PA: American Society for Testing and Materials, 1996).
2. J. G. Quintiere, and M. Harkleroad, *New Concepts for Measuring Flame Spread Properties*, NBSIR 84-2943 (Gaithersburg, MD: National Bureau of Standards, November 1984).
3. J. G. Quintiere, "A Simulation Model for Fire Growth on Materials Subject to a Room-Corner Test," *Fire Safety Journal* 20 (1993): 313–339.
4. P. H. Thomas, "Some Aspects of the Growth and Spread of Fire in the Open," *Forestry* 20, no. 2 (1967): 139–164.
5. P. H. Thomas, "Rates of Spread of Some Wind-driven Fires," *Forestry* 44, no. 2 (1971): 155–175.
6. Y. Nagano, "Fires" in *Comprehensive Study of the Great Hanshin Earthquake*, UNCRD Research Report Series No. 12 (Nagoya, Japan: United Nations Centre for Regional Development, H. Kaji, Director. 1995), 117.
7. K. Akita, "Some Problems of Flame Spread Along a Liquid Surface," *Fourteenth Symposium (International) on Combustion* (Pittsburgh, PA: The Combustion Institute, 1973), 1075–1084.

Burning Rate

Learning Objectives

Upon completion of this chapter, you should be able to:

- Describe the factors influencing energy release rate.
- Execute formulas to predict burning rate and energy release rate.
- Define heat of gasification and heat of combustion.
- Select or construct energy release rate signatures of real items.

INTRODUCTION

Although the fire triangle (fuel, oxygen, and energy) helps to explain the nature of fire; ignition, flame spread, and burning rate, as illustrated in Figure 6-1, comprise the triad of fire growth. Ignition tells us when fire growth begins; flame spread theory allows us to define the extent of the fire's boundaries; and burning rate gives us the consumption of fuel within those boundaries. In this chapter we describe how objects burn, gasifying their liquid or solid fuel from the heat of their own flame. A burning rate theory is presented, but real objects demand real data. We demonstrate how to use these data.

DEFINITION AND THEORY

burning rate, $\dot{m}$
the mass of fuel consumed in the fire per unit time

Burning rate is defined as the mass of solid or liquid fuel consumed per unit time. By *burned*, we mean it reacts with oxygen. In large structural fires, fuel may vaporize but not immediately burn due to limited oxygen. Therefore, one must distinguish between the **mass loss rate** of fuel in fire (fuel supplied) and the mass burning rate (fuel reacted with oxygen). In this chapter, where we are considering the burning of single objects, these two distinctions are considered synonymous. So, if we burn a pan of alcohol on a scale, the heat from the flame will cause the liquid to evaporate and its mass loss can be recorded over time. The mass lost divided by the time to lose this mass is the burning rate, $\dot{m}$, usually expressed in grams per second (g/s).

■ NOTE
It is important to distinguish between mass loss rate of fuel (fuel supplied) and the mass burning rate (fuel reacted with oxygen).

mass loss rate
the mass of fuel *vaporized* but not necessarily burned per unit time

mass burning flux, $\dot{m}''$
burning rate per unit area

In order to ignite the alcohol in the pan in Figure 6-2, it must at least be at its flashpoint temperature. Once a diffusion flame forms over its surface, the surface temperature increases to its boiling point. During this increase, the rate of evaporation or the burning rate also increases accordingly. This changing or unsteady burning rate is typical of the way fuels burn. In general, the rate of burning depends on the fuel properties, its orientation or configuration, and the area involved. The area involved is controlled by ignition and flame spread. The rate of burning over that area may not be steady and may not be uniform. The character of uniformity can be expressed by the quantity: burning rate per unit area or **mass burning flux**, $\dot{m}''$, which describes how each point burns. If we divide this area into small surfaces, associating each surface with a $\dot{m}''$, then the sum of $\dot{m}''$ times each corresponding surface area gives us the total rate of burning.

Figure 6-1 *The triad of fire growth.*

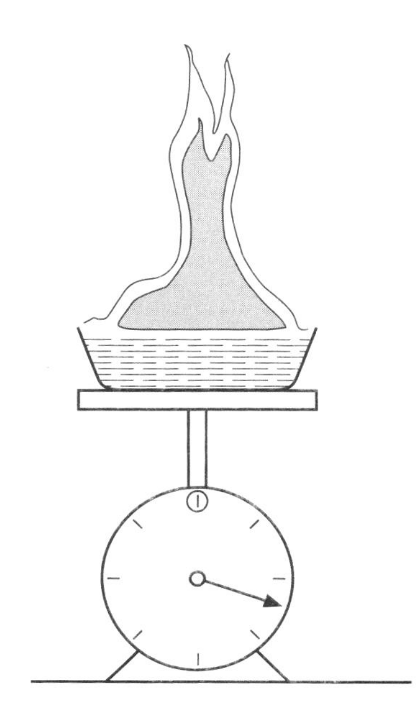

Figure 6-2
Measuring the burning rate.

Heat of Gasification, *L* (kJ/g)

heat of gasification, *L*
energy required to produce fuel vapor from a solid or liquid

reradiation
the radiation reemitted from a heated surface

thermoplastic
a polymer that tends to soften and melt when heated

thermosetting
a polymer with bonds between its chains that tends to char when heated

A general predictive formula for the mass burning rate per unit area, or mass burning rate flux, is given by:

$$\dot{m}'' = \frac{\dot{q}''}{L} \tag{6-1}$$

where $\dot{q}''$ is the net heat flux to the fuel surface, and L is the **heat of gasification**.

The net heat flux principally is due to the flame above the surface, but it can be augmented by other radiant heat sources. These external radiant sources could arise naturally due to a heated room during a fire or due to the configuration of the fuel, as in the interaction of wood logs in a fireplace. Also, these radiant sources could arise in tests for materials to encourage them to burn. The net heat flux is the algebraic sum of the heat flux components (Figure 6-3) at the surface. The **reradiation** term is due to the radiant energy emitted from the surface. For liquids the surface would be at the boiling point. Noncharring **thermoplastic** solids would respond similarly to liquids. But **thermosetting** plastics, wood, and other char formers would exhibit an increasing surface temperature as the char forms an insulating layer. Equation (6-1) is an exact description of the steady burning of liquid fuels and noncharring thermoplastics but is only an estimate of the peak burning

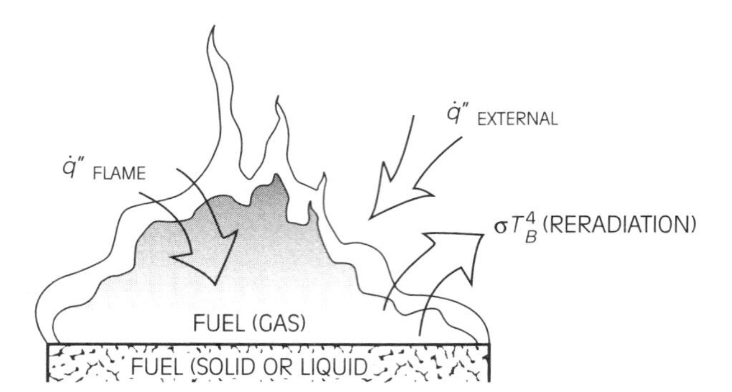

Figure 6-3 *Heat flux components causing burning.*

mass flux for charring materials. For shallow liquid pools or liquid fuels saturated into solids, the formula overestimates the mass burning flux because of heat lost into the substrate of the liquid. Also, for something like gasoline poured on wood, the gasoline would burn off first and even cool the soaked wood in the process.

The quantity L is the energy required to gasify the solid or liquid fuel. It is a thermodynamic property for liquid fuels and can be expressed with great accuracy. For solids, especially charring solids, it is an approximate average property that can vary with time, because the char retains some of the energy. Whereas liquid fuels and noncharring thermoplastics reach steady burning in seconds and minutes respectively, char formers quickly achieve a peak burning flux, then decay over time. This decay is caused by the insulating nature of the char. Later, it is followed by direct oxidation of the char as flaming dies out. In the context of this simple formula, L is only a very approximate property for char formers. Nevertheless, Equation (6-1) gives us a simple, but proper, quantitative tool for estimating the mass burning flux. Unfortunately, the net heat flux $\dot{q}''$, or more specifically, the flame heat flux, is not readily available.

Typical mass burning fluxes range from 5 to 50 g/m^2-s. Values below 5 g/m^{2}s usually lead to extinguishment. As shown directly by Equation (6-1), external heating can enhance the mass burning flux. Such external heating can arise from a fire in a room resulting in a feedback loop that can accelerate burning. Because increasing the atmospheric oxygen leads to an increase in flame temperature, we would expect the resulting flame heat flux to lead to a corresponding increase in mass burning flux. For that reason, burning in elevated oxygen atmospheres will be greater than burning in vitiated atmospheres. Extinguishment, for $\dot{m}''<5$ g/m^2-s, can occur as the oxygen is depleted and can occur if vaporizing water provides a loss in surface heat flux. Surprisingly, there is not a simple theory to explain extinguishment. But the empirical threshold of 5 g/m^2-s is a good benchmark. Suppressing a fire by cooling with water usually leads to extinguishment at this mass flux. Accordingly, Equation (6-1) can help explain a range of burning rate effects.

Typical values of heats of gasification are listed in Table 6-1. These have been

Table 6-1 *Heat of gasification values.*

Fuel	L (kJ/g)
Liquids:	
Gasoline	0.33
Hexane	0.45
Heptane	0.50
Kerosene	0.67
Ethanol	1.00
Methanol	1.23
Thermoplastics	
Polyethylene	1.8–3.6
Polypropylene	2.0–3.1
Polymethylmethacrylate	1.6–2.8
Nylon 6/6	2.4–3.8
Polystyrene foam	1.3–1.9
Flexible polyurethane foam	1.2–2.7
Char Formers	
Polyvinyl chloride	1.7–2.5
Rigid polyurethane foam	1.2–5.3
Whatman filter paper no.3	3.6
Corrugated paper	2.2
Woods	4–6.5

Sources: Data from Tewarson and Quintiere et al., Refs. 1 and 2.

assembled from various sources and are only generically listed. Chemical additives and the physical form of the material (granular, foam, solid) can affect the value of L for solid fuels. Indeed, a chemical retardant could increase L for the solid fuel, making it more difficult to vaporize. As was stated, values of L for pure liquid fuels are precise properties. But, as the degradation process becomes more complex as for char formers, L is an approximate representation of the process. For solid fuels, approximate values of L have been deduced by heating experiments based on using Equation (6-1). In general, it can be seen from Table 6-1 that values of L tend to increase as we move from liquids (<1 kJ/g), thermoplastics (1–3 kJ/g), and char formers (2–6 kJ/g).

Under flaming conditions, liquid fuels attain their boiling point at the surface. Therefore, gasoline would achieve about 30°C at the surface, methanol gets

vaporization temperature the surface temperature required to vaporize a fuel while burning; or in general, the temperature at which a liquid can coexist with its vapor

to 12°C, and kerosene to 232°C. Although the **vaporization temperature** for solid fuels is not a precise property as for pure liquids, typically noncharring thermoplastics achieve surface temperatures of 250–400°C, and char formers reach 400–500°C when burning.

MAXIMUM BURNING FLUX

Although Equation (6-1) is a useful quantitative tool in explaining burning rate, it has limited practical use. The principal problem is knowing how to specify the heat flux term. This quantity can depend on the fuel, its orientation, and its configuration. For burning walls not exceeding 2 m, the wall flame appears to have an approximate constant value of 25±5 kW/m^2. This gives us a strategy for estimating the burning rate of a wall fire. For example, consider a burning wall of nylon 6/6 having a vaporization temperature of 380°C. Assume for the flame: heat flux = 30 kW/m^2; reradiation surface heat flux,

$$\sigma T^4 = 5.67\times10^{-11}\,\frac{\text{kW}}{\text{m}^2\text{K}^4}(380+273)^4$$
$$= 10.3\ \text{kW/m}^2$$

Therefore the net heat flux $\dot{q}'' = 30 - 10.3 = 19.7$ kW/m^2. Because the heat of gasification, L, equals 2.4 kJ/g,

$$\dot{m}'' = \frac{\dot{q}''}{L} = \frac{19.7}{2.4} = 8.2\ \text{g/m}^2\text{-s}$$

This example illustrates how, for this configuration, we can develop an estimate for the burning mass flux.

For other configurations, we can not easily settle on estimates of $\dot{q}''$ without data from specific experimental measurements. Consider the horizontal burning configuration of Figure 6-3. For liquid fuels, a variety of experiments have been conducted. From these experiments it was found that the diameter of the liquid pool is significant. A typical burning mass flux curve is shown in Figure 6-4 for methanol. For very small diameters, $d < 5$ cm, the flame is laminar in character, possessing a very thin flame of nearly 2,000°C. As the diameter increases in this range, the flame sheet moves farther from the surface, reducing the heat flux and consequently the mass flux as shown. For the range of 5–20-cm diameter, the flame heat transfer is more turbulent, with the average turbulent flame temperature becoming approximately 800°C. As we saw in Chapter 3, the convective heat transfer coefficient is roughly 5–10 W/m^2-°C, which yields an approximate constant flame heat flux of 8 kW/m^2 at most. So, for this diameter range, the burning mass flux is nearly constant. But as the diameter further increases, the size and thickness of the turbulent flame increases, providing an increasing source of addi-

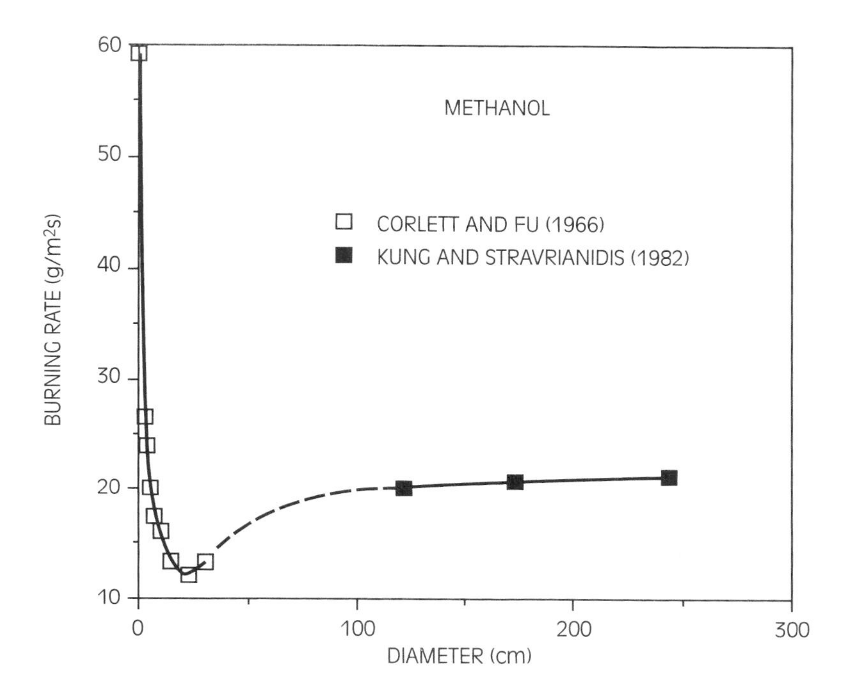

Figure 6-4 *Burning mass flux of methanol pool fires as a function of diameter. From Corlett and Fu; Kung and Stravrianidis, Refs. 3 and 4 respecively.*

tional radiant heat flux from the thick flame. However, as we saw in Chapter 3, the flame radiation reaches a maximum value when the flame emissivity becomes one. As a result, the burning mass flux attains a maximum value. This value depends on the sooting properties of the fuel, but generally occurs for a fire diameter of 1 to 2 m. A similar result is shown for gasoline for diameter values greater than 20 cm in Figure 6-5. The maximum value is seen to reach 55 g/m^{2}s. Such values have been experimentally determined for other liquid fuels, as well as for solid fuel materials. These maximum burning fluxes have been tabulated from various sources in Table 6-2. Although these values must be taken as very rough estimates for the solid fuels, they do offer a framework for computing worst-case burning conditions.

ENERGY RELEASE RATE, $\dot{Q}$

energy release rate, $\dot{Q}$ the energy produced by the fire per unit time, that is, fire power

Perhaps the most important quantity related to fire is the **energy release rate**, the power of the fire measured in kilowatts, kW, and denoted by the symbol, $\dot{Q}$. It, more than any other factor, represents the *size* of the fire and its potential for damage. We shall see that it is directly related to flame height, and from Chapter 3 we

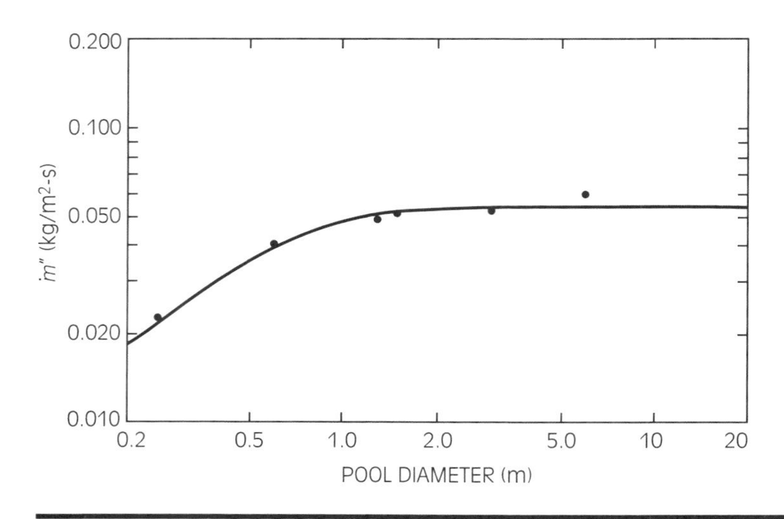

Figure 6-5 *Mass burning flux for gasoline pools. After Ref. 5. Reprinted with permission from the* SFPE Handbook of Fire Protection Engineering. *Copyright © 1995, National Fire Protection Association, Quincy, MA 02269.*

heat of combustion
the energy released by the fire per unit mass of fuel burned

saw that it is directly related to the radiant heat flux surrounding the fire. The entire potential for fire growth and flashover can be related to $\dot{Q}$.

The energy release rate is found through a knowledge of the burning mass flux as where

$$\dot{Q} = \dot{m}'' A \Delta H_c \tag{6-2}$$

■ NOTE
The heat of combustion represents the chemical energy released per unit mass of vaporized fuel that is reacted.

where A is the area involved in vaporization and ΔH_c is the effective **heat of combustion**.

The term *effective* heat of combustion is used as opposed to the *theoretical* heat of combustion because the former applies during the flaming portion of the fire. The heat of combustion represents the chemical energy released per unit mass of vaporized fuel that is reacted. It is possible to measure this value for a solid such as wood in a device called an **oxygen bomb**. In this device all of the combustible content of the wood would be reacted and the resultant energy and mass consumed would be measured. Only inorganic ash would remain. Typically this heat of combustion (theoretical) would be approximately 19 kJ/g. If the same measurements were conducted under typical fire conditions for wood, we would measure ΔH_c of approximately 13 kJ/g for the flaming period and ΔH_c of approxi-

oxygen bomb
a device for measuring the maximum energy released in combustion for a given mass of fuel

Table 6-2 *Maximum burning flux values.*

Fuel	$\dot{m}''$ (g/m²-s)
Liquified propane	100–130
Liquified natural gas	80–100
Benzene	90
Butane	80
Hexane	70–80
Xylene	70
JP-4	50–70
Heptane	65–75
Gasoline	50–60
Acetone	40
Methanol	22
Polystyrene (granular)	38
Polymethyl methacrylate (granular)	28
Polyethylene (granular)	26
Polypropylene (granular)	24
Rigid polyurethane foam	22–25
Flexibie polyurethane foam	21–27
Polyvinyl chloride (granular)	16
Corrugated paper cartons	14
Wood crib	11

Source: From Tewarson, Ref. 1.

mately 30 kJ/g for the smoldering phase of the char. This process is illustrated in Figure 6-6 for a typical woodlike material. Here, we are concerned about the effective heat of combustion during the flaming period. As the char layer builds, flaming will eventually cease at a mass loss flux of 5 g/m^2s or less. Smoldering then consumes the char by oxidation; this process is very slow. In general, both the production of char and the formation of incomplete products of combustion such as soot and carbon monoxide reduces the heat of combustion during flaming. Chemical additives—retardants—can also play a role here. That is why ΔH_c must be measured for a specific material to be accurate.

The effective heats of combustion tend to be highest for gas and liquid fuels, and least for char formers. Table 6-3 lists typical values.

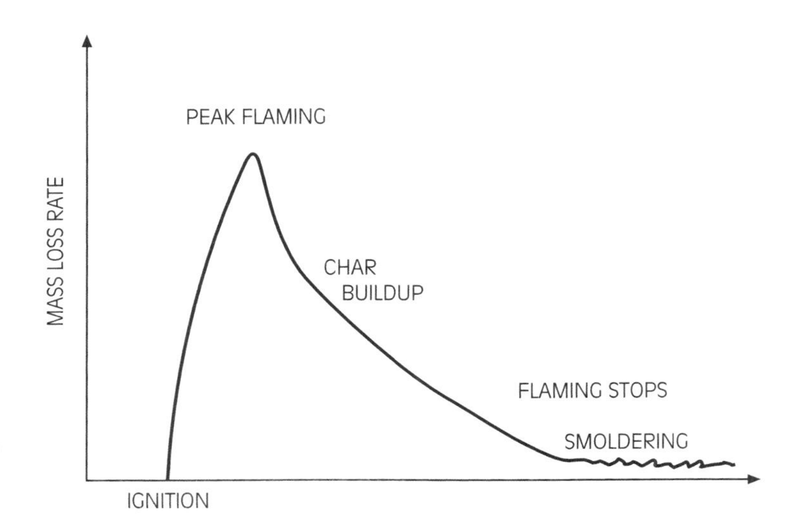

Figure 6-6 *Burning phases for wood.*

NOTE **Liquid fuels are comparably more dangerous than solid fuels.**

An important parameter in assessing fire hazard is the ratio of $\Delta H_c/L$, because this yields the energy released per energy required to vaporize the fuel. It can be seen from Tables 6-1 and 6-3 that liquid fuels are comparably more dangerous than solid fuels, as will be more apparent in the following examples.

Estimating Energy Release Rates

Example: Compute the mass loss rate per unit area and the energy release rate for a 1-m diameter pool fire of wood, polystyrene, heptane, and gasoline. Treat the wood as a crib. Assume the maximum burning rate per unit area applies for these relatively large pool fires. Figure 6-7 depicts the dynamic feature of this pool fire example.

1. Wood. From Table 6-2 for wood cribs, $\dot{m}'' = 11$ g/m^2-s. (Note: If this were actual wood cribs or stacked pallets, we would have to consider the exposed surface. Instead, here we consider only a top surface representing a shallow array.)

$$A = \frac{\pi}{4}D^2 = \frac{\pi}{4}(1\text{ m})^2 = 0.785\text{ m}^2$$

$$\begin{aligned}\dot{Q} &= \dot{m}'' A \Delta H_c \\ &= (11\text{ g/m}^2\text{s})\,(0.785\text{m}^2)\,(15.0\text{ kJ/g}) \\ &= 130\text{ kJ/s} = 130\text{ kW}\end{aligned}$$

Table 6-3 *Effective heat of combustion, ΔH_c (kJ/g).*

Methane	50.0
Ethane	47.5
Ethene	50.4
Propane	46.5
Carbon monoxide	10.1
n-Butane	45.7
c-Hexane	43.8
Heptane	44.6
Gasoline	43.7
Kerosene	43.2
Benzene	40.0
Acetone	30.8
Ethanol	26.8
Methanol	19.8
Polyethylene	43.3
Polypropylene	43.3
Polystyrene	39.8
Polycarbonate	29.7
Nylon 6/6	29.6
Polymethyl methacrylate	24.9
Polyvinyl chloride	16.4
Cellulose	16.1
Glucose	15.4
Wood	13–15

Source: Based on data from Tewarson, Ref. 1.

2. Polystyrene. Table 6-2 gives for a granular fuel

$$\dot{m}'' = 38 \text{ g/m}^2\text{-s}$$

$$\begin{aligned}\dot{Q} &= \dot{m}'' A \Delta H_c \\ &= (38 \text{ g/m}^2\text{-s})\,(0.785 \text{ m}^2)\,(39.85 \text{ kJ/g}) \\ &= 1189 \text{ kW}\end{aligned}$$

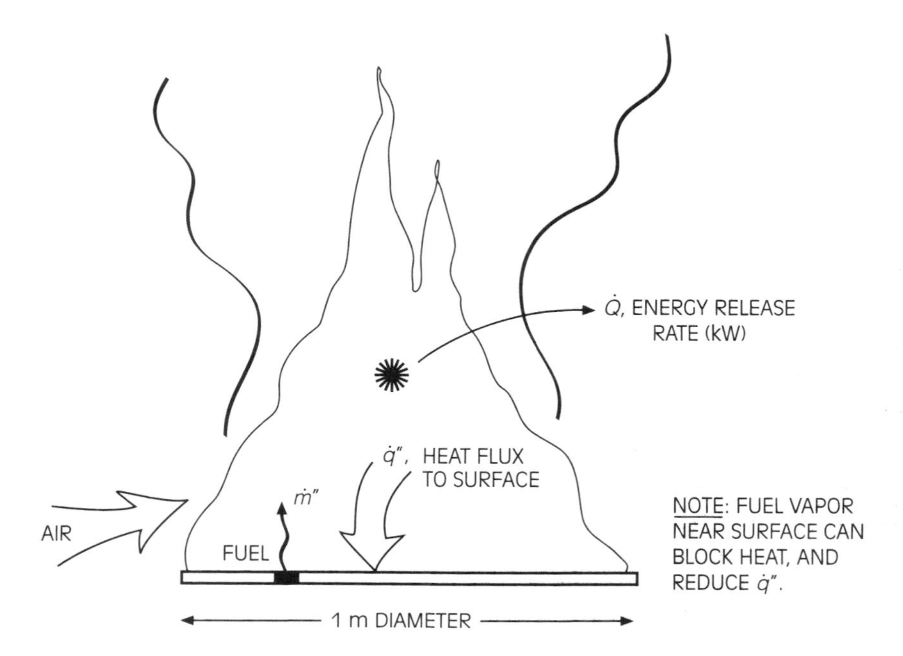

Figure 6-7 *Example configuration.*

3. Heptane. Table 6-2 gives 65 to 75 g/m²-s for $\dot{m}''$. Therefore, we see some differences that must depend on the source of the data. We will use the higher value, 75 g/m²-s.

$$\dot{Q} = (75\ \text{g/m}^2\text{-s})\ (0.785\ \text{m}^2)\ (44.6\ \text{kJ/g})$$
$$= 2626\ \text{kW}$$

4. Gasoline. Again, using Table 6-2 we find 55 g/m²-s.

$$\dot{Q} = \dot{m}'' A \Delta H_c$$
$$= (55\ \text{g/m}^2\text{-s})\ (0.785\ \text{m}^2)\ (43.7\ \text{kJ/g})$$
$$= 1887\ \text{kW}$$

Summary for 1-m diameter pool fires.

	$\dot{m}''$ (g/m²-s)	$\dot{Q}$ (kW)
Wood	11.	130.
Polystyrene	38.	1189.
Heptane	75.	2650.
Gasoline	55.	1887.

These results are very illustrative of the relative potential for damage of these four fuels. Despite their relatively small area (1-m diameter), the liquid fuels would present fires that can rapidly bring typical residential rooms to full involvement. The polystyrene would also pose a serious threat, but the wood at this size would be a manageable fire. However, a net heat flux of 18 kW/m^2 is causing the gasoline fire compared to 65 kW/m^2 for the polystyrene fire. These values can be computed from Equation (6-1) and Table 6-1. The difference in these heat fluxes accentuates our lack of understanding in how to predict them from theory. Their difference is almost counterintuitive, because the gasoline fire is larger in power, $\dot{Q}$. One explanation suggests these differences are due to radiation absorption. The large amount of vaporized gasoline by-products near the fuel surface acts like a cloud in blocking the flame radiation.

EXPERIMENTAL RESULTS FOR SELECTED ITEMS

The only practical way to determine the burning rate or energy release rate of an item is by direct measurement. This measurement depends on the manner of igni-

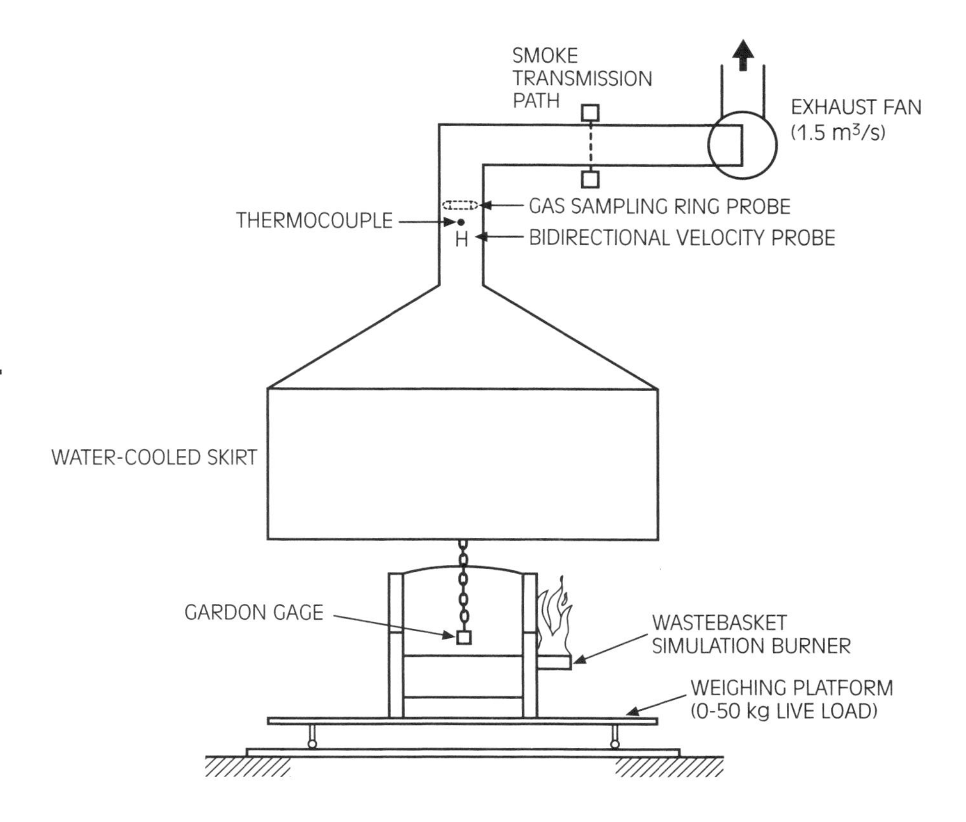

Figure 6-8 *The furniture calorimeter. After Ref. 5. Reprinted with permission from the* SFPE Handbook of Fire Protection Engineering. *Copyright © 1995, National Fire Protection Association, Quincy, MA 02269.*

■ NOTE
The only practical way to determine the burning rate or energy release rate of an item is by direct measurement.

oxygen consumption calorimeter
device to measure energy release rate

burnout
point at which flames cease

tion, particularly for the initial fire growth over time. Weighing devices or load cells can be used to determine mass loss while an item is burning. If the products of combustion are collected in an exhaust duct and the oxygen content of these gases are measured, the energy release rate can also be determined. Such a device is called an **oxygen consumption calorimeter** (Figure 6-8). The device works on the principle that the heat of combustion per unit mass of oxygen consumed is nearly a constant (13 kJ/g) for a wide range of ordinary fuel compounds. The energy release rate is found directly from recording the rate of oxygen consumed. Examples of results for upholstered furniture are shown in Figure 6-9. Typically, the early part of the growth curve depends on the ignition process. The steep increase in $\dot{Q}$ is due to the overall fire growth—flame spread and burning rate—for the furniture. The decay from the peak energy release rate reflects the **burnout** of various portions of the furniture. The width of the curves, or more precisely, the area under them, represents the amount of burnable fuel.

The simplest way to estimate energy release is to measure mass loss rate by burning an object on a scale. From Equation (6-2) the $\dot{Q}$ is found by multiplying this mass loss rate by an appropriate heat of combustion. Table 6-4 gives some typical peak values of burning rate for real items.

Figures 6-10 through Figure 6-20 give experimental results for a range of items, including trash, televisions, Christmas trees, upholstered chairs, and pallets.

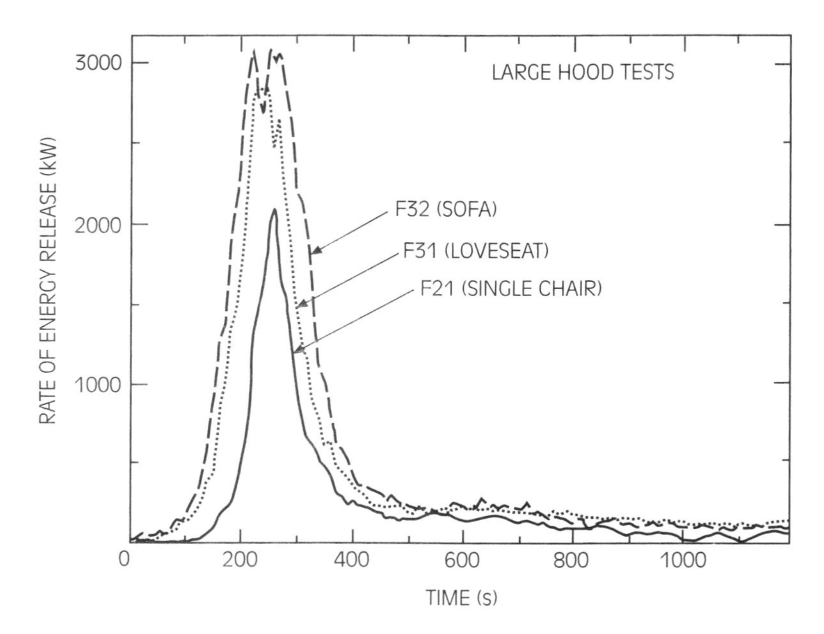

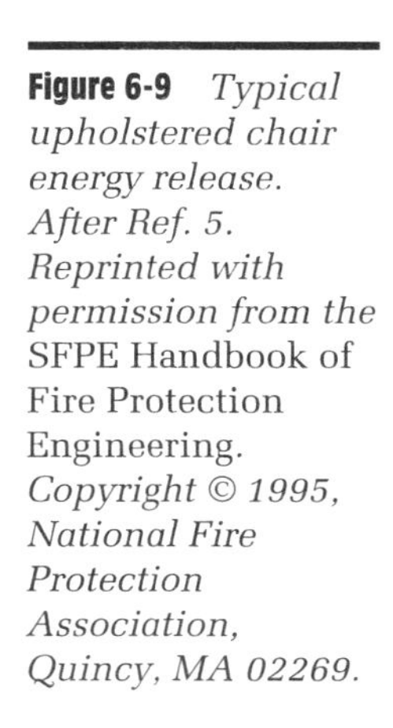
Figure 6-9 *Typical upholstered chair energy release. After Ref. 5. Reprinted with permission from the* SFPE Handbook of Fire Protection Engineering. *Copyright © 1995, National Fire Protection Association, Quincy, MA 02269.*

Table 6-4 *Typical peak burning rate values (in g/s).*

Small waste containers (18–40ℓ)	3–6
Large waste containers (70–120ℓ)	5–10
Chairs, wood and upholstered	10–60
Sofas	20–100
Beds	20–140
Closet	~40
Office	~90
Bedroom	~130
Kitchen	~190
House	~30,000

Source: Based on data from Ref. 6.

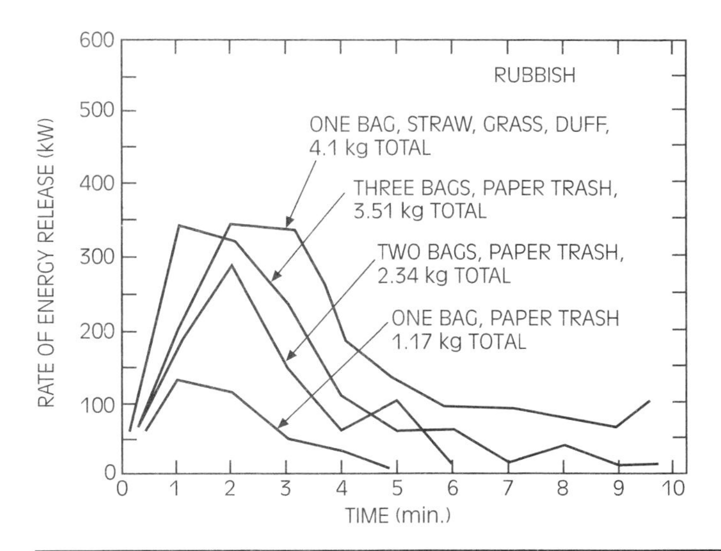

Figure 6-10 *Energy release rate of representative rubbish. After Ref. 5. Reprinted with permission from the* SFPE Handbook of Fire Protection Engineering. *Copyright © 1995, National Fire Protection Association, Quincy, MA 02269.*

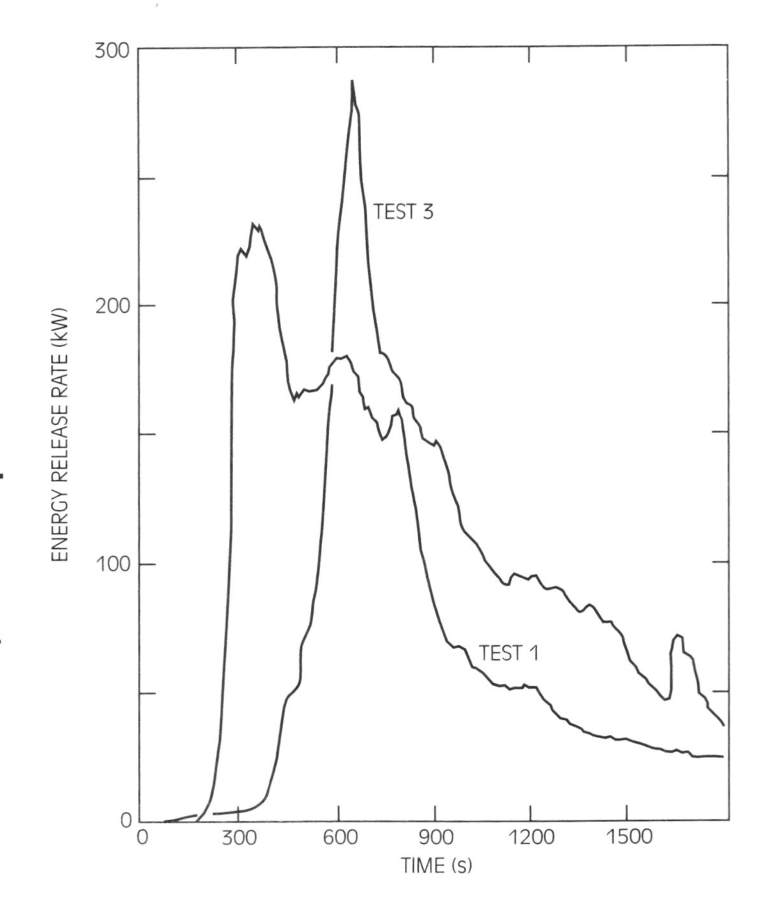

Figure 6-11 *Energy release rate results for television sets. After Ref. 5. Reprinted with permission from the* SFPE Handbook of Fire Protection Engineering. *Copyright © 1995, National Fire Protection Association, Quincy, MA 02269.*

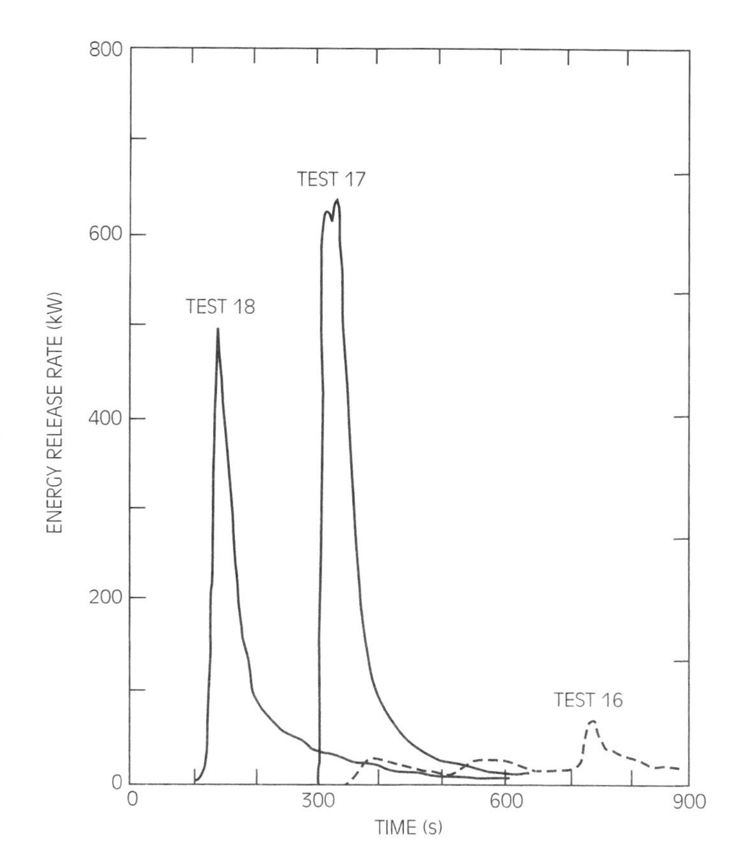

Figure 6-12 *Energy release rate results for Christmas trees. After Ref. 5. Reprinted with permission from the* SFPE Handbook of Fire Protection Engineering. *Copyright © 1995, National Fire Protection Association, Quincy, MA 02269.*

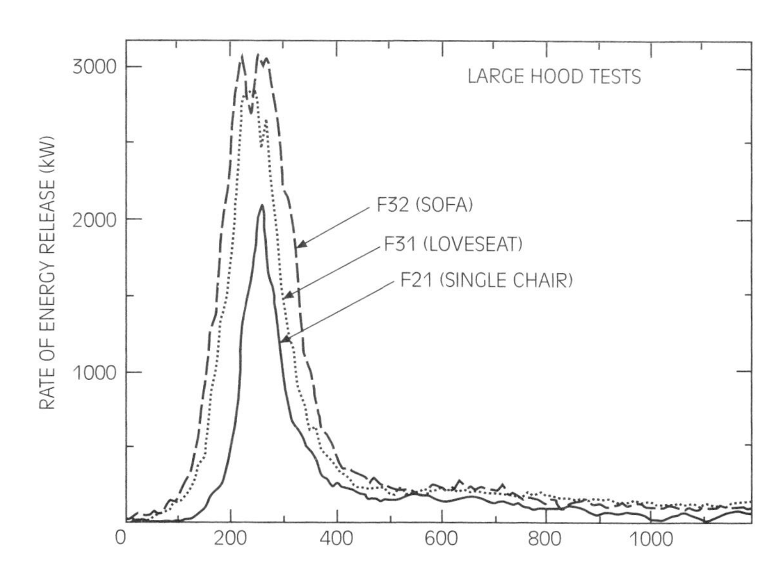

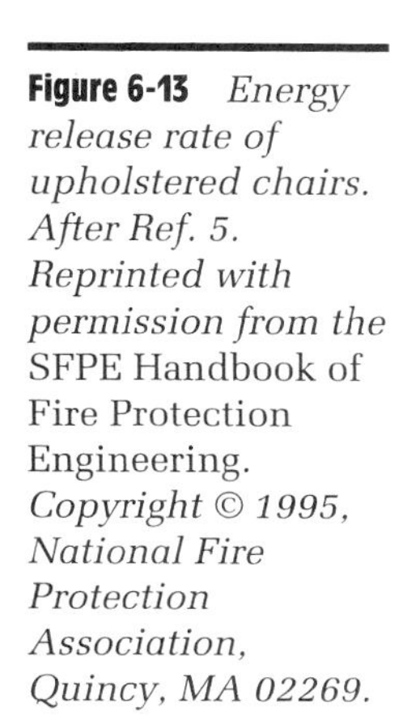

Figure 6-13 *Energy release rate of upholstered chairs. After Ref. 5. Reprinted with permission from the* SFPE Handbook of Fire Protection Engineering. *Copyright © 1995, National Fire Protection Association, Quincy, MA 02269.*

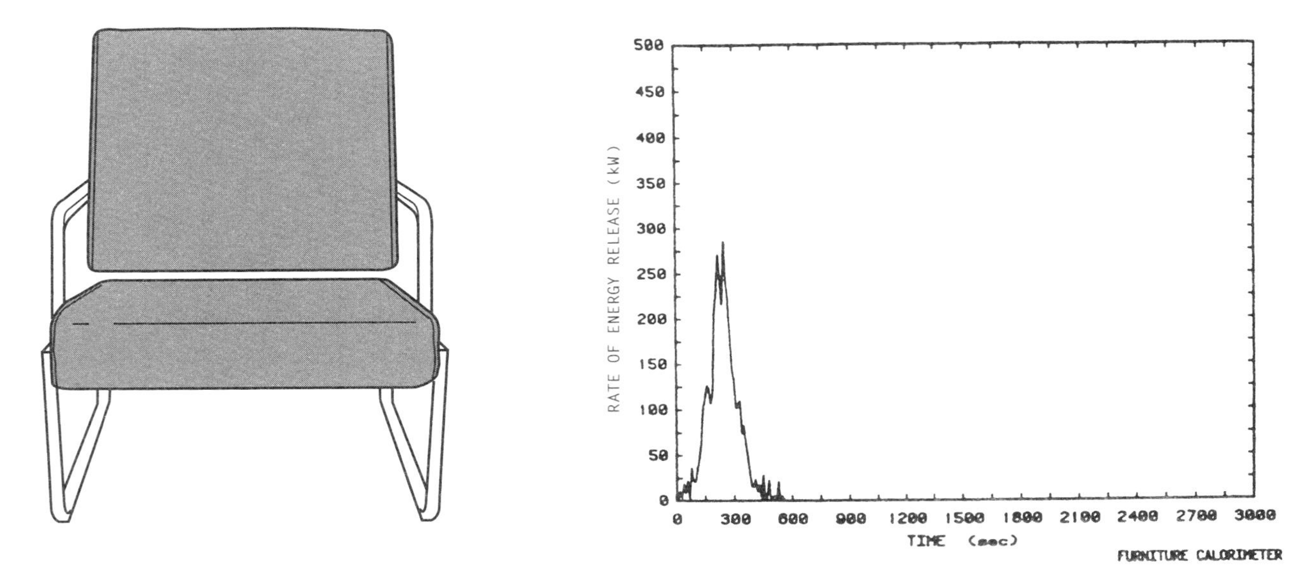

Figure 6-14 *Upholstered chair. After Gross, Ref. 8.*

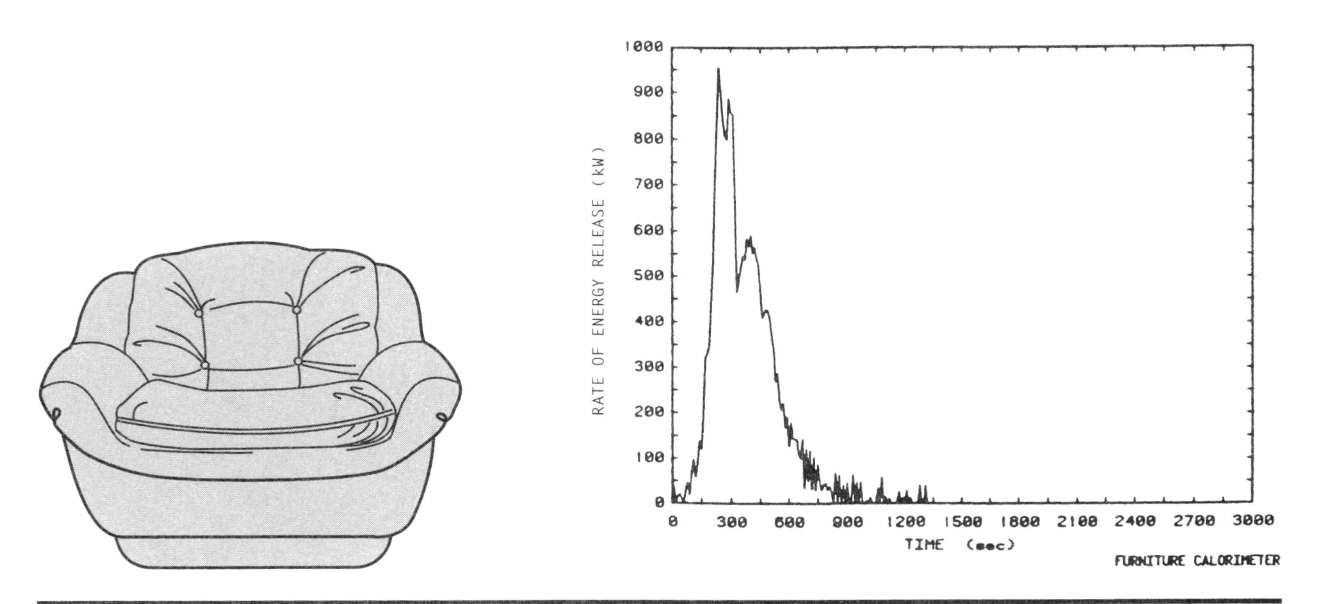

Figure 6-15 *Upholstered chair. After Gross, Ref. 8.*

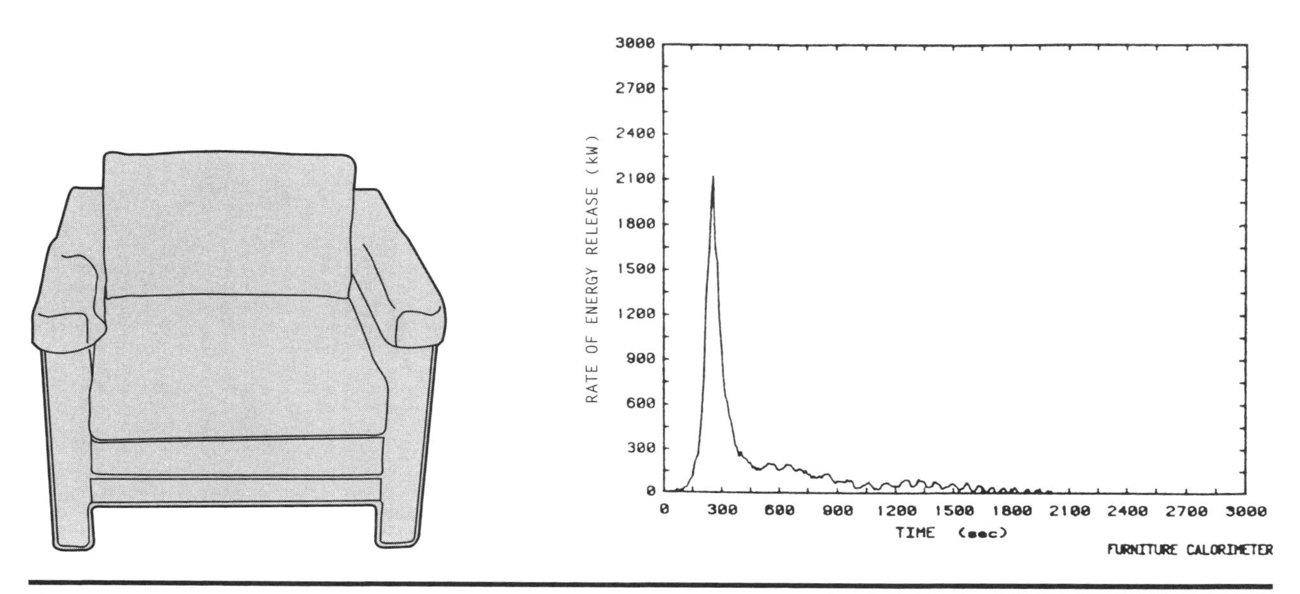

Figure 6-16 *Upholstered chair. After Gross, Ref. 8.*

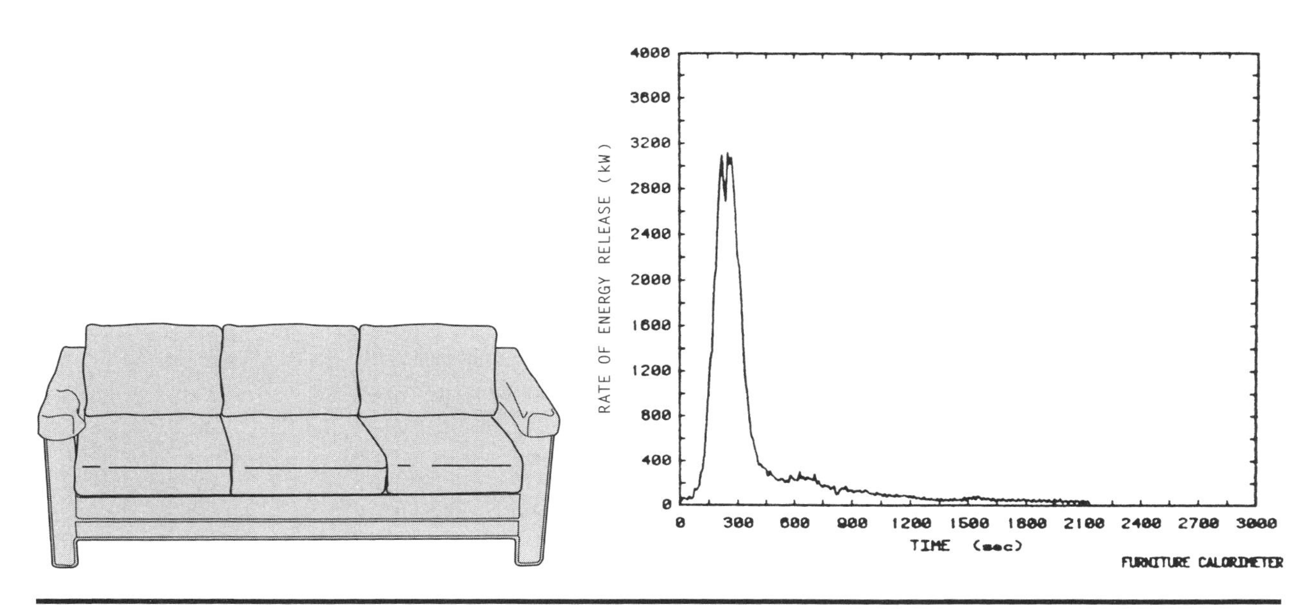

Figure 6-17 *Sofa. After Gross, Ref. 8.*

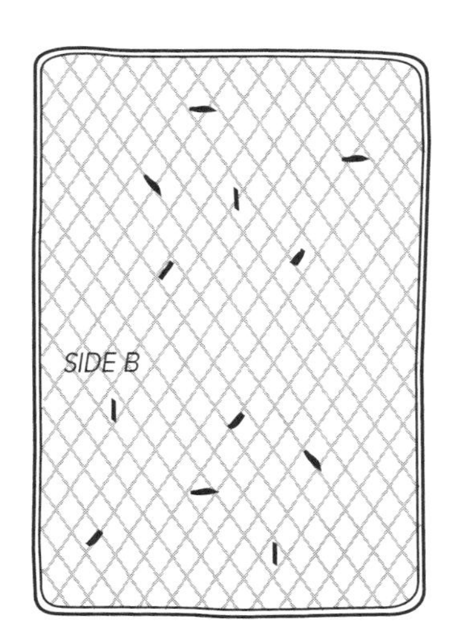

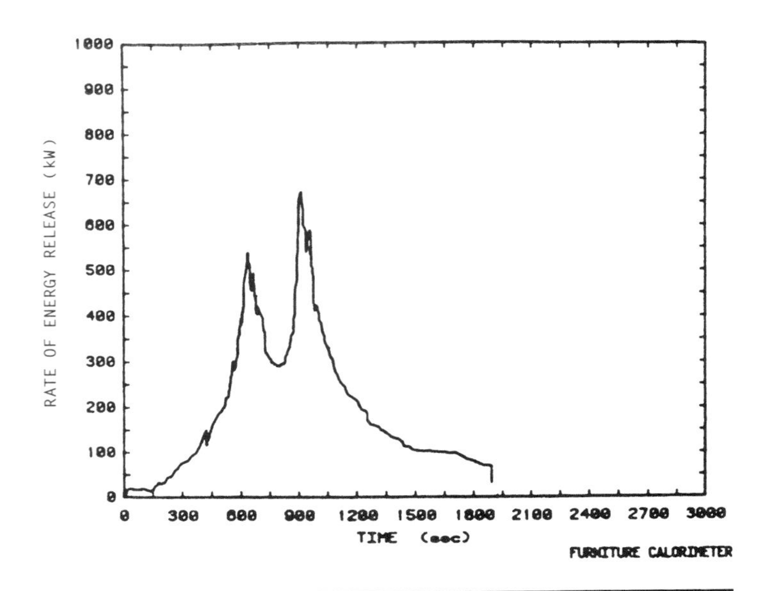

Figure 6-18 *Mattress. After Gross, Ref. 8.*

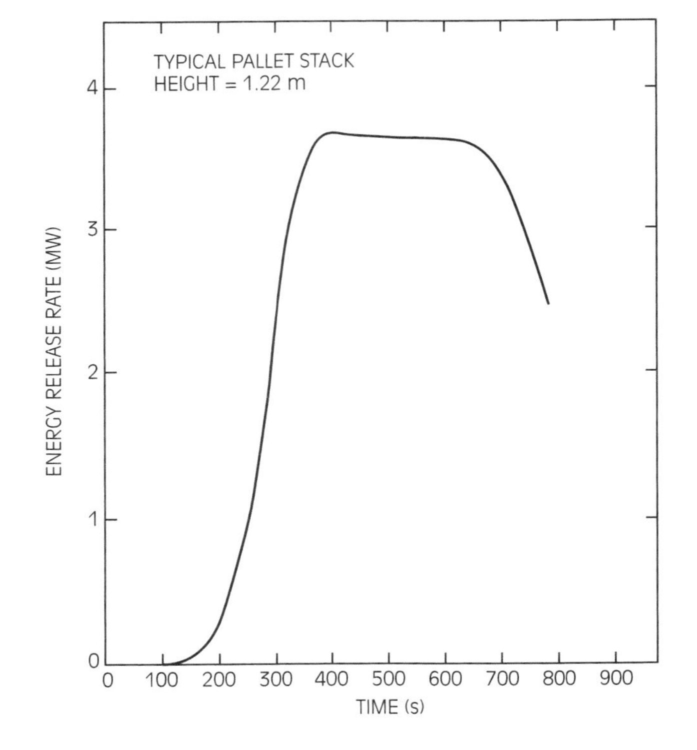

Figure 6-19 *Typical energy release rate for a wood pallet stack. After Ref. 5. Reprinted with permission from the* SFPE Handbook of Fire Protection Engineering. *Copyright © 1995, National Fire Protection Association, Quincy, MA 02269.*

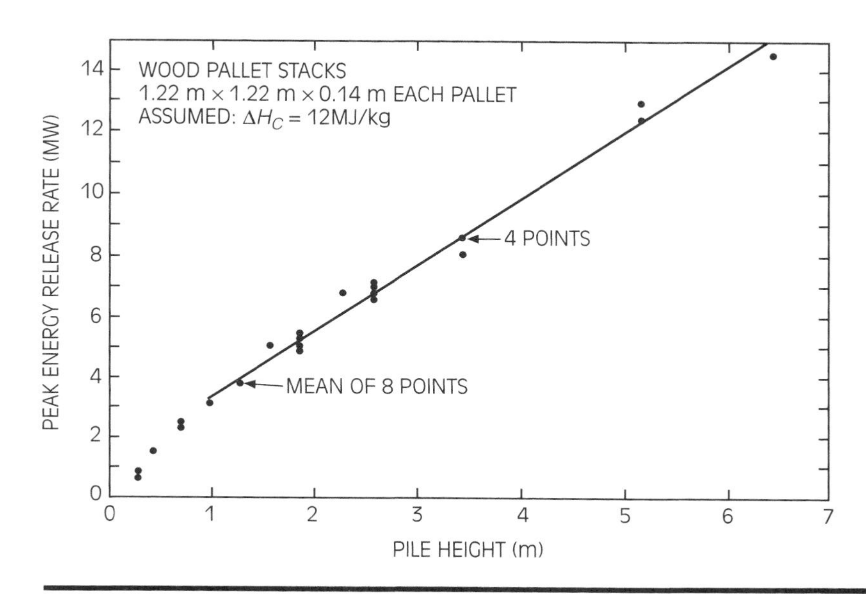

Figure 6-20 *Dependence of pallet burning rate on stack height. After Ref. 5. Reprinted with permission from the* SFPE Handbook of Fire Protection Engineering. *Copyright © 1995, National Fire Protection Association, Quincy, MA 02269.*

These results show the variation in burning behavior within the category of a single generic item. This variation can be due to many factors: size, materials, nature of the ignition source, and the effect of fire retardants. As with Table 6-4, they represent a range of typical values. To accurately know how an item will burn, it must be tested in a device similar to that shown in Figure 6-8. Table 6-5 gives typical experimental values for warehouse commodities in terms of floor area covered. These values should be interpreted as peak values for a fixed floor area without any lateral spread.

FIRE GROWTH RATE

As we have seen, the fire growth rate depends on the ignition process; flame spread, which defines its perimeter; and the mass burning flux over the area involved. For items of furniture and commodities, this complex process cannot be predicted by simple formulas. However, each item, due to its composition and configuration, must have a characteristic growth time. In other words, a given item, once ignited, may achieve 1 MW (Megawatt, a million watts) in 130 seconds. For

Table 6-5 *Fire behavior of warehouse commodities: fully involved energy release rates for a fixed floor area.*

Commodity	$\dot{Q}$/Floor Area Covered (MW/m²)
Methanol	0.72
Diesel oil	1.9
Kerosene	2.2
Gasoline	2.2
Wood pallets, stacked 1 1/2 ft high	1.3
Wood pallets, stacked 5 ft high	3.7
Wood pallets, stacked 10 ft high	6.6
Wood pallets, stacked 15 ft high	9.9
Mail bags, filled, stored 5 ft high	0.39
PE letter trays, filled, stacked 5 ft high	8.2
PS insulation board, rigid foam, stacked 14 ft high	3.1
PU insulation board, rigid foam, stacked 15 ft high	1.9
PS tubs rested in cartons, stacked 14 ft high	5.1
FRP shower stalls in cartons, stacked 15 ft high	1.2
PE bottles in cartons, stacked 15 ft high	1.9
PS toy parts in cartons, stacked 15 ft high	2.0
PE trash barrels in cartons, stacked 15 ft high	2.9
Cartons, compartmented, stacked 15 ft high	2.2
PVC bottles packed in cartons, compartmented, stacked 15 ft high	3.3
PP tubs packed in cartons, compartmented, stacked 15 ft high	4.2
PE bottles packed in cartons, compartmented, stacked 15 ft high	6.1
PS jars packed in cartons, compartmented, stacked 15 ft high	14.0

Source: Based on data from Heskestad, Ref. 7.

PE = polyethylene, PU = polyurethane, PVC = polyvinyl chloride, PS = polystyrene, PP = polypropylene, FRP = fiberglass-reinforced-polyester

another object, it might take 80 seconds. Such data have been compiled and are useful in fire analyses. Also it has been found that the growth rate approximately follows a relationship proportional to time squared. An explanation for this is that the flame spread velocity is constant over time. If the rate of fire growth depends on the fire itself, the growth rate would be exponential (e^t), not t^2. Nevertheless, an empirical growth rate formula is

$$\dot{Q} = at^2 \quad \text{(6-3)}$$

where a is a constant associated with the item. If t_1 is the characteristic time to reach 1 MW, then $a = 1\ \text{MW}/t_1^2$. This growth rate expression has been fitted to experimental data by ignoring the early incubation period where the growth process is establishing a foothold. This fit is illustrated in Figure 6-21.

Table 6-6 lists fire growth times (t_1) corresponding to t^2 fires for commodities that have unlimited lateral extent for growth. In reality, at some point the decay process must be described as indicated by the burning curves shown in Figures 6-10 through 6-19. The t^2 relationship has proved useful and has been adopted into NFPA 72B Appendix B [9] to categorize fires for detector analysis and into NFPA 92B [10] for design of smoke control systems. These NFPA standards classify the growth times as follows and as illustrated in Figure 6-22:

slow	$t_1 = 600$ s
medium	$t_1 = 300$ s
fast	$t_1 = 150$ s
ultrafast	$t_1 = 75$ s

A practical application to explain the entire growth and decay process could combine the data of Tables 6-5 and 6-6. The early growth can be considered as t^2 up to the full involvement rate of Table 6-5 for the floor area covered. Burning ceases when the available mass of fuel is depleted at the rate specified.

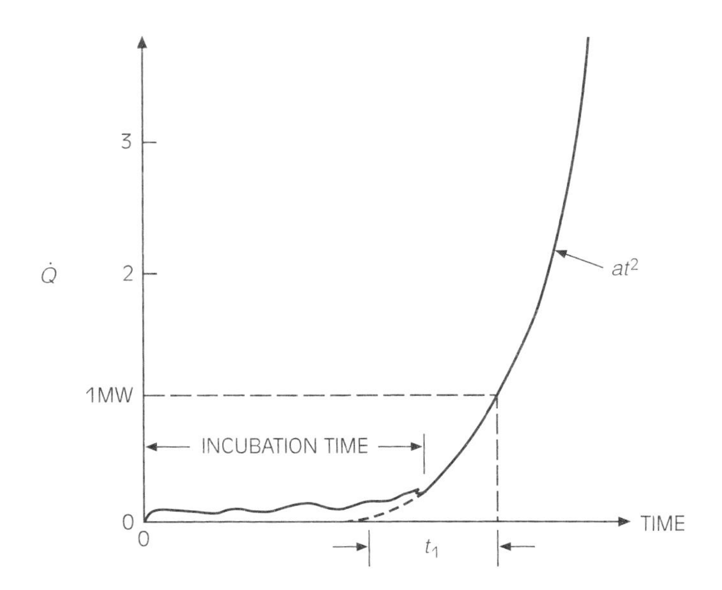

Figure 6-21 *Fire growth curve of t^2 fitted to data. After Heskestad, Ref. 7.*

Table 6-6 *Characteristic times to reach 1 MW for t^2 fires.*

Commodity	t_1 (s)
Wood pallets, stacked 1 1/2 ft high	155–310
Wood pallets, stacked 5 ft high	92–187
Wood pallets, stacked 10 ft high	77–115
Wood pallets, stacked 16 ft high	72–115
Mail bags, filled, stored 5 ft high	187
Cartons, compartmented, stacked 15 ft high	58
Paper, vertical rolls, stacked 20 ft high	16–26
Cotton, polyester garments in 12 ft high rack	21–42
"Ordinary combustibles" rack storage, 15–30 ft high	39–262
Paper products, densely packed in cartons, rack storage, 20 ft high	461
PE letter trays, filled, stacked 5 ft high on cart	189
PE trash barrels in cartons, stacked 15 ft high	53
PE bottles packed in compartmented cartons, 15 ft high	82
PE bottles in cartons, stacked 15 ft high	72
PE pallets, stacked 3 ft high	145
PE pallets, stacked 6–8 ft high	31–55
PU mattress, single, horizontal	115
PU insulation board, rigid foam, stacked 15 ft high	7
PS jars packed in compartmented cartons, 15 ft high	53
PS tubs nested in cartons, stacked 15 ft high	115
PS insulation board, rigid foam, stacked 14 ft high	6
PUS bottles packed in compartmented cartons, 15 ft high	8
PP tubs packed in compartmented cartons, 15 ft high	9
PP and PE film in rolls, stacked 14 ft high	38
Distilled spirits in barrels, stacked 20 ft high	24–39

Source: Based on data from Heskestad, Ref. 7.

PE = polyethylene, PU = polyurethane, PVC = polyvinyl chloride, PS = polystyrene, PP = polypropylene, FRP = fiberglass-reinforced-polyester

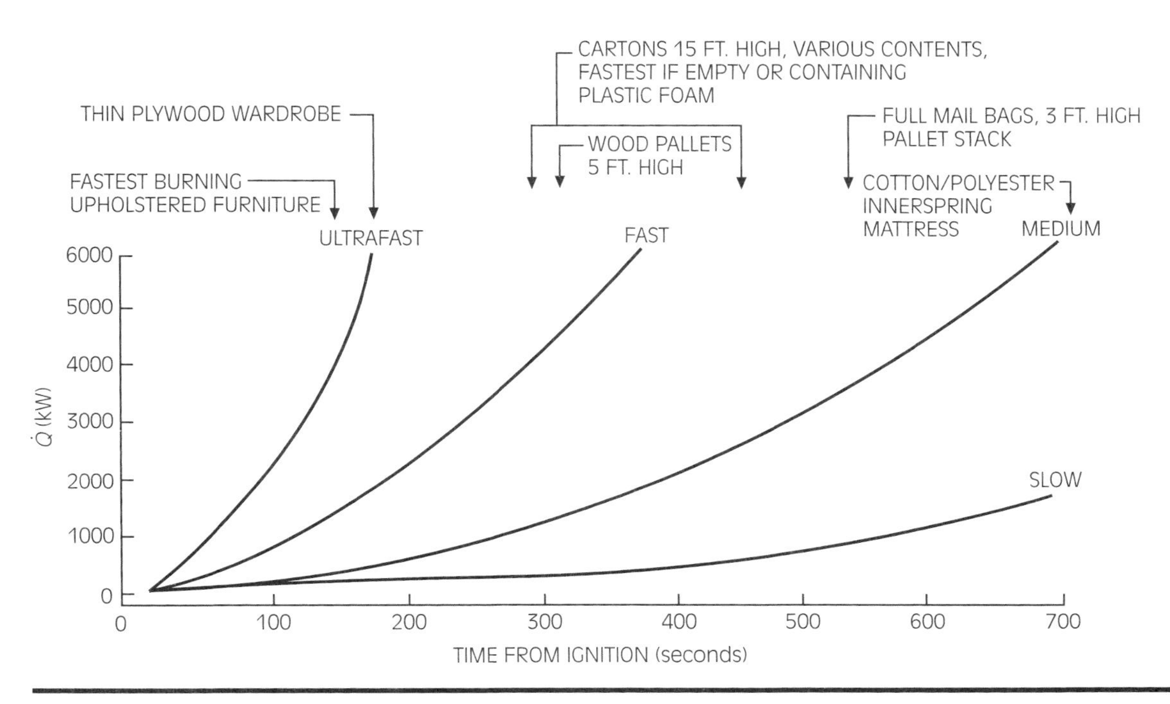

Figure 6-22 *Categories of t^2 fires. After Ref. 9. Reprinted with permission from 1995 National Fire Codes, Vol.10, NFPA* 92B, 1991 ed. *National Fire Protection Association, Quincy, MA 02269. The reprinted material is not the official position of the National Fire Protection Association, which is represented by the standard in its entirety.*

Summary

After ignition a fire may grow over an area by flame spread. The burning rate (or fuel supply rate) can be determined for a material or item by knowing its mass burning flux ($\dot{m}''$), usually given in g/m^{2}s. Values less than 5 usually lead to extinguishment, and typical values range from 5 to 50 g/m^2-s for surfaces, and higher for stacked commodities. The property called *heat of gasification* measures how much energy it takes to produce this gasified fuel; the property called *heat of combustion* gives how much energy is released when it burns.

A simple theory was presented to compute the mass burning flux from surfaces from its heat of gasification provided the flame heat flux is known. Despite some data, a prediction of flame heat flux for a given fuel is beyond current science. However, much data exists on the maximum mass burning flux for fuels and for actual items burned. Energy release rate data, derived from large-scale calorimeters, provide a source for making estimates. The energy release rate (or alternatively, the burning rate) is the most significant index of fire hazard. It relates to the potential for ignition of nearby items, to flashover potential in a room, and to the rate of water needed to extinguish the fire.

Review Questions

1. Plexiglas, polymethyl methacrylate (PMMA), is often used as display windows and cases in art galleries and museums. If its flame imparts 30 kW/m^2 to the surface when it burns, estimate the energy release rate of a 2 m^2 square sheet (one side). You will need Tables 6-1 and 6-3 on pages 105 and 111 respectively. Assume its vaporization temperature is 350°C.
2. Wood pallets 5 feet high and 6 × 6 ft in their plan dimension are ignited and burn. Using Tables 6-5 and 6-6 on pages 122 and 124 respectively, estimate the following:
   a. Maximum energy release rate.
   b. Time to reach 1 MW.
   c. Its t^2 energy release rate curve.

True or False

1. All the gaseous fuel produced in a fully involved room fire burns in the room.
2. Burning rate can be determined by weighing the object as it burns.
3. The heat of gasification and the heat of combustion are usually equal.
4. Fires always grow proportional to time squared.
5. The heat flux from a turbulent flame, like its temperature, is nearly the same for all fuels.

Activities

1. Obtain small shallow glass dishes ranging in size from 1 cm to 10 cm in diameter. Pour about 1/4-in. depth of a liquid fuel of high flashpoint (> 25°C) into a dish. Ignite it and time its burning period. How does burning time vary with fuel diameter? Explain your results.
2. Obtain a collection of small wood sticks, roughly 4 × 1/4 × 1/4 inch. Make a pile about 4 × 4 × 2 high, first putting all the sticks in contact, then leaving 1/4 inch spacing between the sticks in alternating layers. Try to ignite both with a match flame. Which pile burns? Why?
3. Examine the items contained in a room. From the data of this chapter, try to estimate the potential energy release rate of each item that could burn. Add to find their sum. Recognize that in a typical residential room, 1 MW could jeopardize it into flashover.

References

1. A. Tewarson, "Generation of Heat and Chemical Compounds in Fires," chap. 3–4 in *SFPE Handbook of Fire Protection Engineering*, 2d ed., edited by P. J. DiNenno, (Quincy, MA: National Fire Protection Association, June 1995).
2. J. G. Quintiere, G. Haynes, and B. T. Rhodes, "Applications of a Model to Predict Flame Spread over Interior Finish Materials in a Compartment," *Journal of Fire Protection Engineering* 7, no. 1 (1995): 1–14.
3. R. C. Corlett, and T. M. Fu, "Some Recent Experiments with Pool Fires," *Pyrodynamics* 4 (1966): 253–269.
4. H. C. Kung, and P. Stravrianidis, "Buoyant Plumes of Large-Scale Pool Fires," in *Nineteenth Symposium (International) on Combustion* (Pittsburgh, PA: The Combustion Institute, 1982), 905–912.
5. V. Babrauskas, "Burning Rates," chap. 3-1 in *SFPE Handbook of Fire Protection Engineering*, 2d ed., edited by P. J. DiNenno (Quincy, MA: National Fire Protection Association, June 1995).
6. J. Quintiere, "Growth of Fire in Building Compartments," in *Fire Standards and Safety*, ASTM STP 614, edited by A. F. Robertson (West Conshohocken, PA: American Society for Testing and Materials, 1977), 131–167.
7. G. Heskestad, "Venting Practices," sec. 6, chap. 10 in *Fire Protection Handbook*, 17th ed., edited by A. E. Cote and J. L. Linville (Quincy, MA: National Fire Protection Association, 1991), 6.104–6.116.
8. D. Gross, *Data Sources for Parameters Used in Predictive Modeling of Fire Growth and Smoke Spread*, NBSIR 85-3223 (Gaithersburg, MD: National Bureau of Standards, September 1985).
9. "Guide for Smoke Management Systems in Malls, Atria, and Large Areas," *National Fire Codes*, Vol. 10, NFPA 92B, 1991 ed., (Quincy, MA: National Fire Protection Association, 1995), 92B–30.

Fire Plumes

Learning Objectives

Upon completion of this chapter, you should be able to:

- Calculate flame height.
- Estimate temperature above a fire.
- Describe the behavior of fire plumes.
- Describe the meaning of buoyancy.

INTRODUCTION

All fires involve buoyancy, which sets the flow pattern and the nature of the flame. This flow pattern directs the growth of the fire and sucks in the air supply. It is also responsible for the characteristics of the smoke that departs the fire. Even under wind conditions, buoyancy plays a significant role in the fire's behavior. In this chapter, we only consider the symmetric fire on a horizontal surface (floor). This is commonly called a "pool" fire, representative of a liquid fuel spill. We do not address wind effects. However, mild wind effects (1 m/s) can significantly alter the orientation of flames, and rotational flows can cause fire whirls (rotating columns of elongated flames). For the pool fire, we examine the height of its flame and the temperatures of its gases.

TURBULENT FIRE PLUMES

fire plume
the flame and gases emanating from a burning object

buoyancy
an effective force on fluid due to density or temperature differences in a gravitational field

A **fire plume** is the buoyant column of flame and hot combustion products rising above the fuel source. **Buoyancy** is a force that arises in a fluid due to density differences. Because density is inversely proportional to temperature for gases, the higher than air fire plume temperatures cause an upward force on the hot gases relative to the surrounding air. (See Figure 7-1.) If the hot plume gases cool to the local air temperature, the buoyant force becomes zero, causing the plume to cease rising. This condition is responsible for cigarette smoke stratifying in smoke-filled rooms and for increases in atmospheric pollution due to temperature inversions.

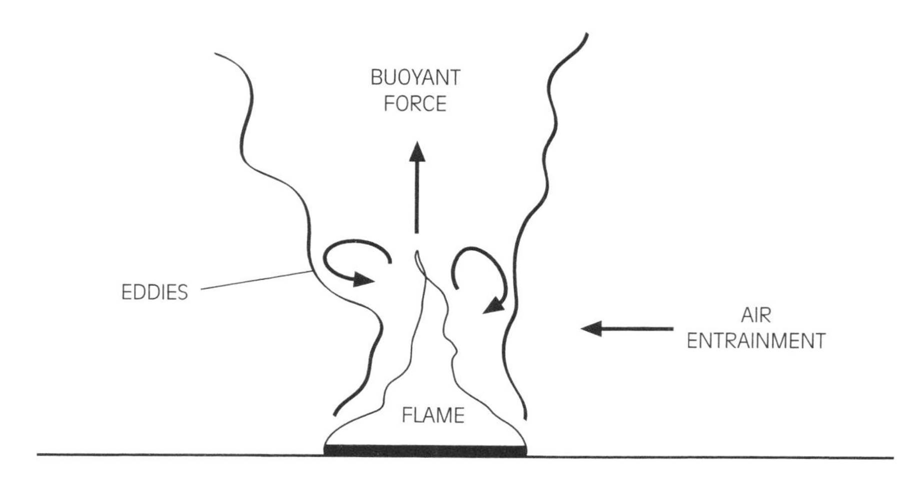

Figure 7-1 *Buoyancy and fire plumes.*

entrainment
the process of air or gases being drawn into a fire, plume, or jet

flame height
the vertical measure of the combustion region

eddies
rotating regions of a fluid

vortex
a ring of eddies

As the hot gases rise in a buoyant plume, the colder air is induced to flow into the fire plume. This flow process is called **entrainment** (see Figure 7-1). The rate of this entrainment flow is responsible for the **flame height** and the characteristics of the fire plume. Consequently, the fire is intertwined with the air flow, each depending on the other.

Flow can be laminar or turbulent. Laminar flow is smooth and orderly. The candle flame is laminar, but its buoyant plume of heated gases becomes turbulent at roughly 1 foot (or less) above the flame. This characteristic can be seen by projecting the candle flame on a screen with a collimated beam of light shining through it. The fluctuating light and dark images of the plume illustrate high frequency turbulence. Within a turbulent flame this high frequency motion of the reaction zone can produce oscillations of hundreds of degrees at a fixed point. Superimposed on these high frequency fluctuations, as high as 10 cycles per second (10 Hertz, HZ), are large-scale fluctuations as a direct result of buoyancy and fluid friction. As the buoyancy causes the plume fluid to rise, its cooler edges are slowed and fall back down. This results in **eddies**, comparable to the size of the fuel diameter, as illustrated in Figure 7-2. These eddies rotate as well as rise along the edges of the plume. They undoubtedly influence entrainment.

The large eddy turbulent effects present in all buoyant plumes can have a profound effect on flame height in combusting plumes. Although these turbulent flames are diffusion flames, the diffusion mechanism of bringing the fuel and oxygen together is *riding* on the turbulent flow. The outer eddies produce a rising **vortex** between them. The high frequency turbulence distorts this image, as illustrated in Figure 7-3. As this burning eddy rises, it extends the flame until the fuel burns out. A new eddy then carries the flame to this former height. As a result, the flame can fluctuate from a maximum to a minimum. Figure 7-3 shows that the maximum height can be roughly twice its minimum. Although shown for a 0.3-m diameter burner with an eddy frequency of 3 Hz, this difference in heights is typical of buoyant turbulent flames. When we address flame height and plume temperatures, we should keep in mind that the flame is unsteady, undergoing high frequency and the slower eddy fluctuations. Formulas for flame height usually only give average results and do not account for the periodic "ripping off" of the

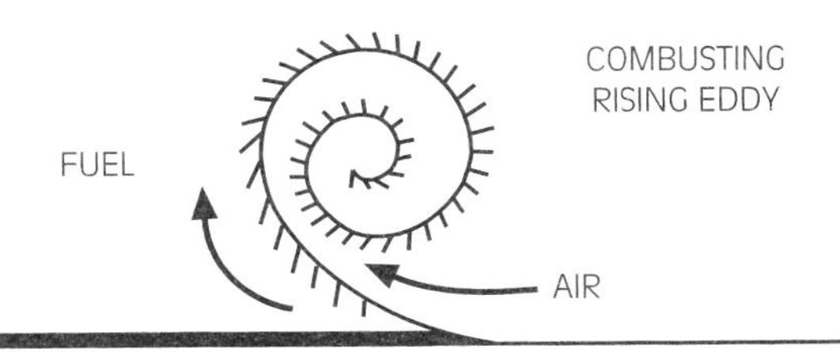

Figure 7-2 *Turbulent burning eddy.*

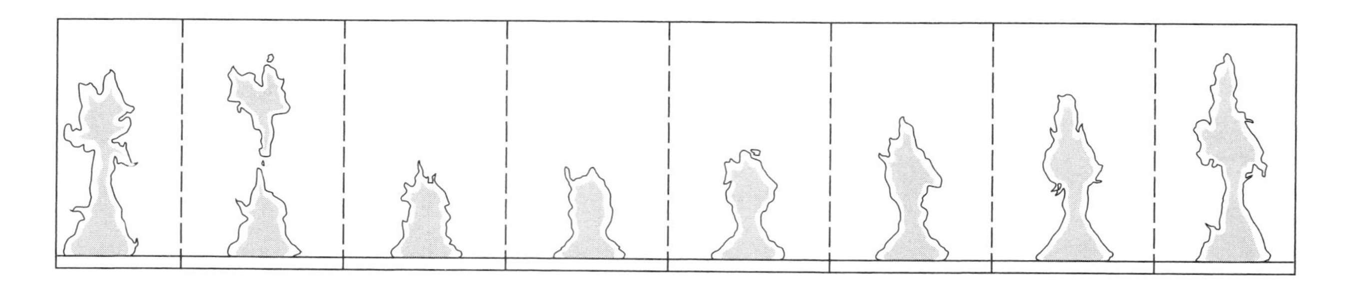

Figure 7-3 *A sequence of flame images over approximately 0.3 seconds for a 0.3-diameter fire. After McCaffrey, Ref. 1.*

flame shown in Figure 7-3. Also the frequency of formation and departure of the outer eddies from the base of the plume (called *vortex shedding*) depends on the base diameter of the fire, D. This vortex shedding frequency in Hertz for a fire of diameter D in meters is approximately

$$f = 1.5/\sqrt{D} \tag{7-1}$$

A 10-cm fire would shed eddies about 5 times per second, but a large wildland fire of 100 m would form a vortex every 10 s. Figure 7-4 shows a photograph of a large liquid fuel fire (approximately 10-m diameter) that illustrates the eddies. Note that the eddies are of size comparable to the diameter, whereas the high frequency turbulence has much finer length scales. This photograph also illustrates the engulfment of the flame by black soot carried by these eddies to the colder fringes of the flame where the soot does not burn.

FLAME HEIGHT

pool fires
fires involving horizontal fuel surfaces, usually symmetrical

line fires
elongated fires on a horizontal fuel surface

From here on throughout the chapter, we restrict our attention to symmetrical fires from circular horizontal fuel sources. These are called **pool fires**, representative of liquid fuel spills. However, the results can approximate other symmetric fuel shapes burning on a floor. **Line fires** and wall fires have somewhat different quantitative characteristics, but are qualitatively similar. We also address principally turbulent fires with no wind effects and dominated by buoyant effects. However, let us first consider a range of conditions to give a quantitative perspective.

Jet Flames

Consider gaseous fuel issuing from a pipe of diameter D at an exit velocity V_e. For a given fuel and a fixed pipe diameter, the ignited gases will result in an increas-

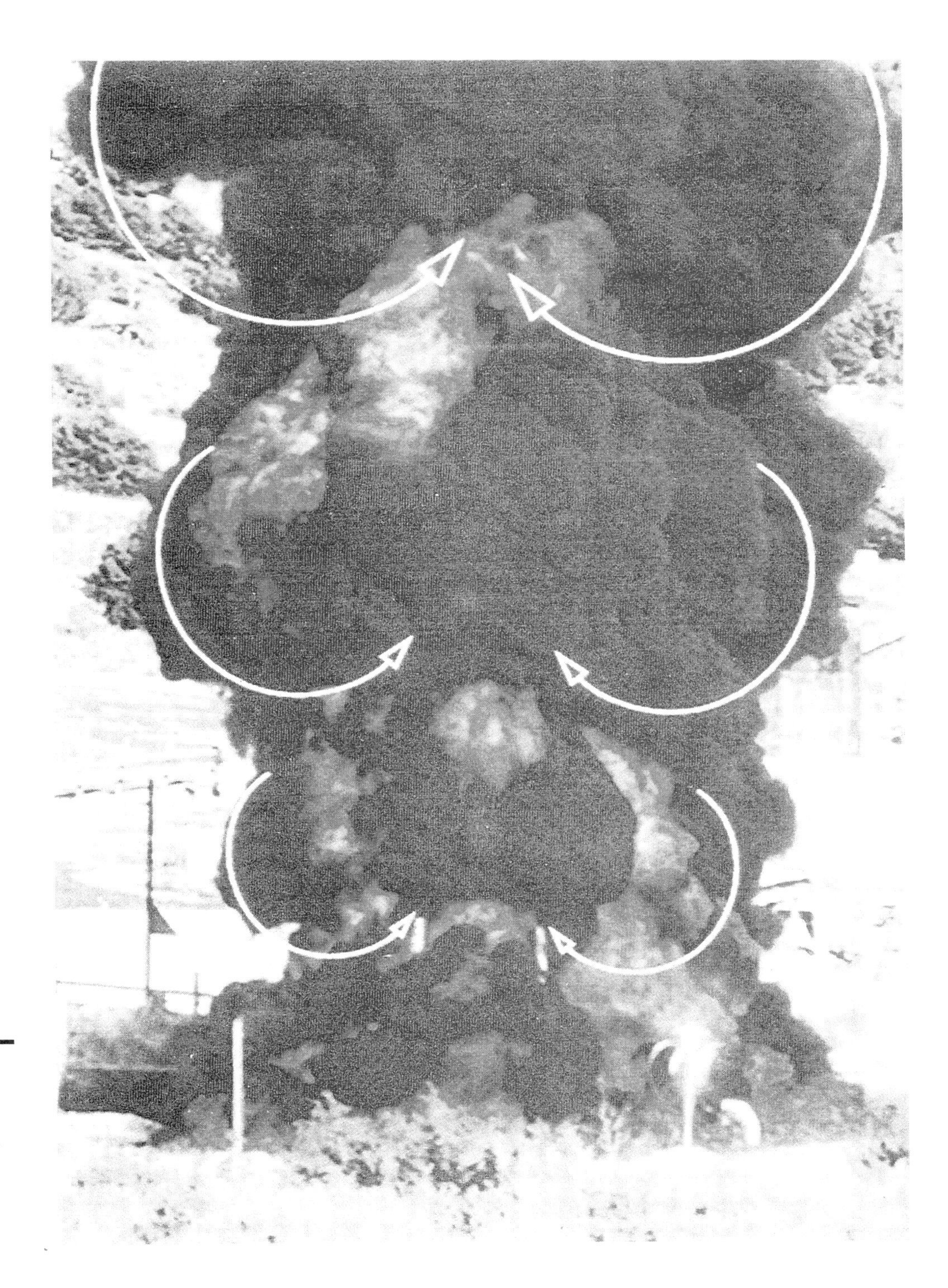

Figure 7-4
Photograph of a large-scale liquid fuel fire (D ~ 10m). Courtesy of L. A. Gritzo, Sandia National Laboratories.

ing laminar flame as V_e is increased, as shown in Figure 7-5. As the exit velocity is increased the flame is maintained as laminar up to a maximum length of nearly a distance of 200 diameters. At higher velocities, the flow becomes turbulent and the flame length becomes fixed for a given fuel and pipe diameter. This phenomenon is a direct result of the entrainment process. The basis for the prediction of flame length is that the flame ends when all of the fuel is burned. Because the fuel is supplied at the base, the end of the combustion process is controlled by how fast the air is entrained over the length of the fire plume. This situation is illustrated in Figure 7-6.

For the turbulent fuel jet of Figure 7-5, it can be shown that

$$\text{rate of air mass entrained} \sim \rho_a D L_f V_e \tag{7-2}$$

where ρ_a is the density of the surrounding air,
D is the pipe exit flow diameter,
and L_f is the flame length.

The rate of fuel supplied is

$$\dot{m}_f = \rho_f V_e \frac{\pi D^2}{4} \tag{7-3}$$

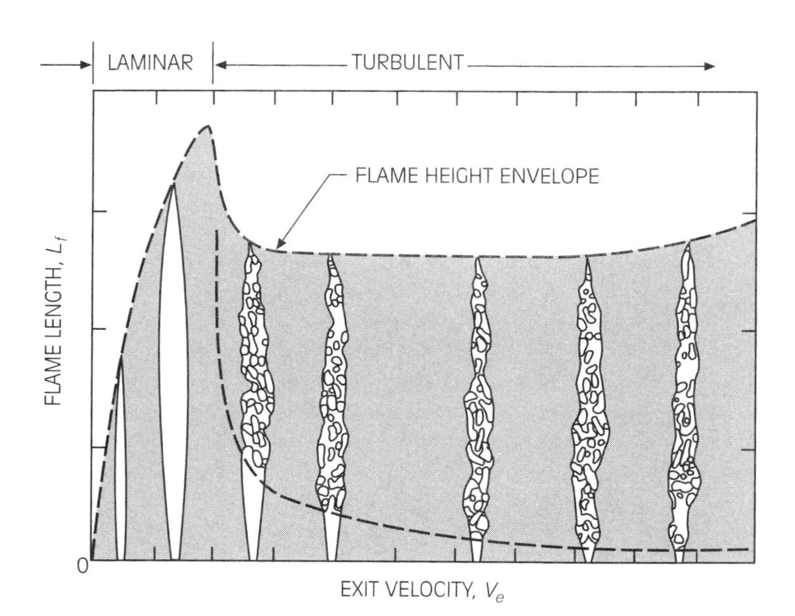

Figure 7-5 *Flame lengths for gaseous jets issuing from a fixed diameter pipe. After Hottel and Hawthorne, Ref. 2.*

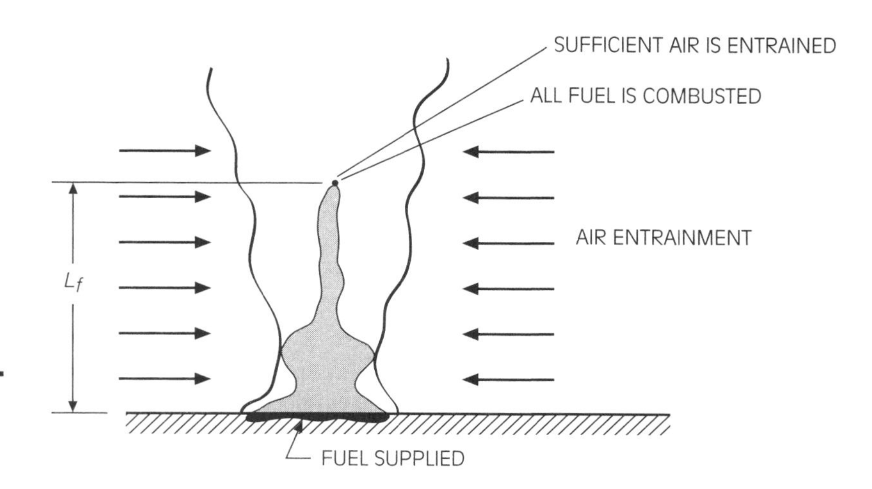

Figure 7-6 *Air entrainment and flame length.*

stoichiometric
refers to the proportional amount of air needed to burn the fuel

fuel lean
description of fuel burning in an excess supply of air

stoichiometric air to fuel mass ratio, *s*
the ratio of air to fuel by mass needed to burn all the fuel to completion

where ρ_f is the gaseous fuel density at the pipe's exit. These two mass flow rates must be at least in **stoichiometric** proportion is order for the air to burn all of the fuel. Usually, for turbulent conditions, this process may not be so efficient and it requires more air than needed for stoichiometric combustion. That is, we will have more air than needed, or the process is said to be **fuel lean**. Recall that stoichiometric here implies that the **stoichiometric air to fuel mass ratio** (s) is that determined for a complete chemical reaction. Therefore, from Equations (7-2) and (7-3), we have the proportionality:

$$\frac{\rho_a D L_f V_e}{\rho_f \frac{\pi}{4} D^2 V_e}$$

or solving,

$$\frac{L_f}{D} \sim \frac{(\rho_f)s}{\rho_a} \qquad \textbf{(7-4)}$$

This result explains why the turbulent jet flame length becomes constant for a given fuel (having density ρ_f and stoichiometric ratio s) at a given diameter. Such turbulent jet flames commence at exit velocities much higher than those due to buoyancy-induced flame velocities. They would be relevant for ruptures of high pressure gas lines. We shall more clearly quantify this distinction in the next section.

Pool Fire Flames

If we examine the "exit" velocity of fuel gases leaving the vaporizing solid or liquid fuel, we would find these are very low, not in the range to be considered *jet* velocities. For example, in Chapter 6 we saw that the maximum burning mass flux for a gasoline pool fire is 55 g/m^2-s. Because the vaporized gasoline is about twice as dense as air, the velocity of these vapors would be approximately

$$V_e \approx \frac{55 \text{g}/\text{m}^2 - \text{s}}{2000 \text{ g}/\text{m}^3} \approx 3 \text{ cm/s}$$

This velocity is *not* responsible for the subsequent entrainment of air. That entrainment results from the substantial increase in fire plume velocities due to combustion and its resulting buoyancy.

Let us estimate this buoyant flow velocity. A simple explanation for this velocity is that the relative buoyant potential energy is converted into kinetic energy. For a unit volume of gas in the fire plume, Figure 7-7 illustrates this energy transfer. We reason that the relative potential energy per unit volume at height z under the gravitational acceleration g is

$$(\rho_a - \rho)\, gz$$

and the kinetic energy per unit volume (considering negligible starting kinetic energy) is

$$\rho \frac{V^2}{2}$$

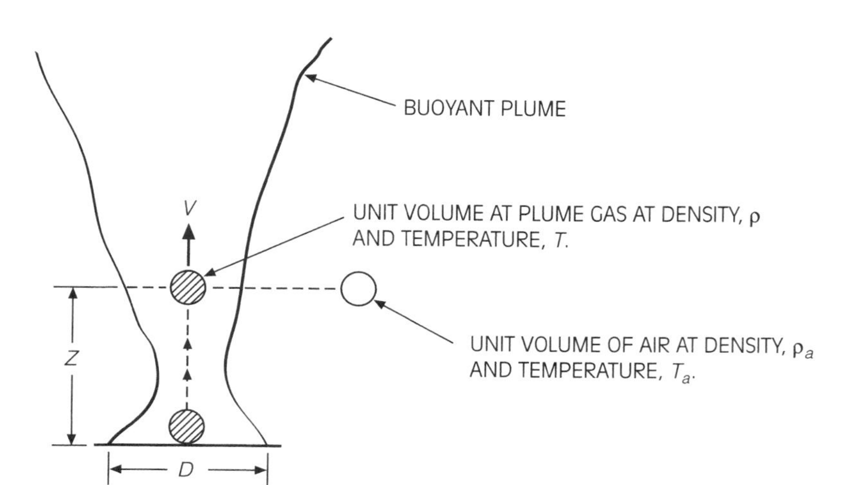

Figure 7-7 *Exchange of buoyant potential energy into kinetic energy in the fire plume.*

Equating these energies gives the plume velocity as

$$V = \sqrt{\frac{2(\rho_a - \rho)gz}{\rho}}$$

or

$$V = \sqrt{\frac{2(T - T_a)gz}{T_a}} \quad \text{(7-5)}$$

where $T/T_a = \rho_a/\rho$, i.e., the temperature is inversely dependent on density. Because the gravitational acceleration, g, is 9.81 m/s^2, this velocity is approximately 4.5 m/s for $z = 1$ m and $(T–T_a)/T_a = 1$. So, for our gasoline fire example with a flame height of 1 m, the fire plume velocity is more than 100 times the vaporization velocity (i.e., 3 cm/s). This fire plume velocity controls the entrainment of air. By a process similar to that which led to Equation (7-4), except with the air entrainment velocity proportional to the plume velocity of Equation (7-5), we can obtain a formula for the height of this buoyant flame. It can be shown that

$$\frac{L_f}{D} \sim Q^{*2/5}$$

dimensionless
having no units of measure (terms combine to produce no units)

This quantity Q^* is a **dimensionless** property of the turbulent fire, which is computed as

$$Q^* = \frac{\dot{Q}}{\rho_a c_{pa} T_a \sqrt{gD}\, D^2} \quad \text{(7-6)}$$

where ρ_a is the air density (1.2 kg/m^3)
c_{pa} is the air specific heat (1.0 kJ/kgK)
T_a is the air temperature (293K)
$\dot{Q}$ is the energy release rate of the fire (kW)
D is the pool diameter (m), and
g is the gravitational acceleration (9.81 m/s^2).

In these units

$$Q^* = \frac{\dot{Q}\,(\text{kW})}{1101\,[D(\text{m})]^{5/2}} \quad \text{(7-7)}$$

Our 1-m diameter, gasoline fire of Chapter 6 where $\dot{Q} = 1887$ kW has a Q^* of 1.7. Typical natural fires may range from Q^* values of 0.5 to 100 for very large diameter fires to stacked wooden pallets, respectively.

Equation (7-6) suggests that Q^* is the ratio of combustion energy to a nominal flow energy of the plume. However, by taking the 2/5 power,

$$Q^{*2/5} = \frac{(\dot{Q}/\rho_a c_{pa} T_a \sqrt{g})^{2/5}}{D} \quad \textbf{(7-8)}$$

characteristic combustion length, $Q^{*2/5}$ a length scale representative of the fire size

the numerator now has dimensions of length. This numerator represents a **characteristic combustion length** of the fire and is directly related to the flame length.

Figure 7-8 shows a range of experimental results that shows the full spectrum from low Q^* fires (conflagrations) to jet flames where buoyancy is unimportant ($\sim Q^* = 10^6$). An approximate demarcation between purely buoyant pool fire behavior and where jet flames commence is $Q^* \sim 10^4$. Over a range of most natural fires $0.5 \leq Q^* < 100$, a relatively simple formula (from Heskestad[4]) gives a good estimate of the average flame height.

$$L_f = 0.23\ \dot{Q}^{2/5} - 1.02\ D \quad \textbf{(7-9)}$$

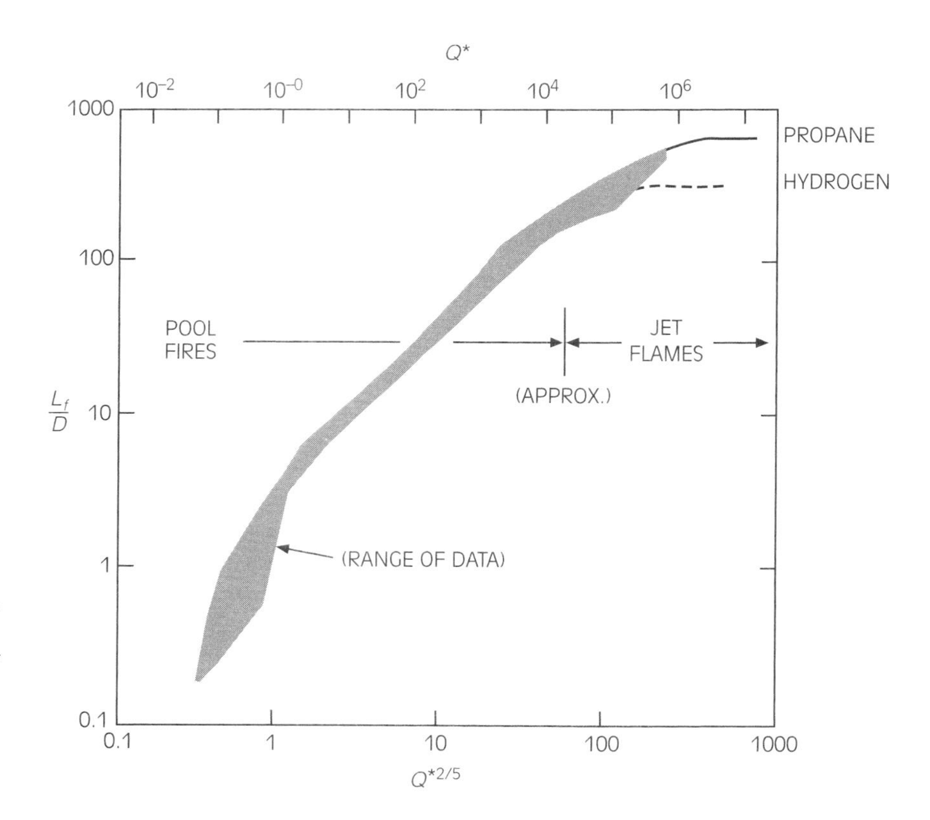

Figure 7-8 *Flame length for a range of symmetrical fire conditions in terms of Q^*. After McCaffrey, Ref. 3.*

where $\dot{Q}$ is in kilowatts and L_f and D are in meters. Recall that the flame height may fluctuate by a factor of 2, so Equation (7-9) reflects a degree of uncertainly at most. The most uncertainty exists below Q^* of 0.5. We shall see from the next section that Equation (7-9) favors the maximum flame height.

FIRE PLUME TEMPERATURES

The temperatures within a turbulent fire plume vary across the width of the plume from a maximum at the center to the air temperature at the edge. We shall consider the center line maximum temperature along the height (z) of a turbulent buoyant fire plume. The variation with height depends on whether z is within the combustion zone ($z \leq L_f$) or the noncombusting zone ($z > L_f$). We can consider a simple theory to explain these vertical temperature variations. Over the combusting region, the flame generates $\dot{Q}$ energy release rate with X_r, the fraction of the energy radiated away from the flame. This energy radiates away from the flame, reaching a maximum of $(1\text{-}X_r)\dot{Q}$ at the end of the combustion region at $z = L_f$. The energy release rate is distributed over the flame length as the entrained air reacts with the fuel present. Once at the fuel is reacted at $z = L_f$, the energy rate $(1\text{-}X_r)\dot{Q}$ remains constant after that throughout the noncombusting plume region. These processes are illustrated in Figure 7-9.

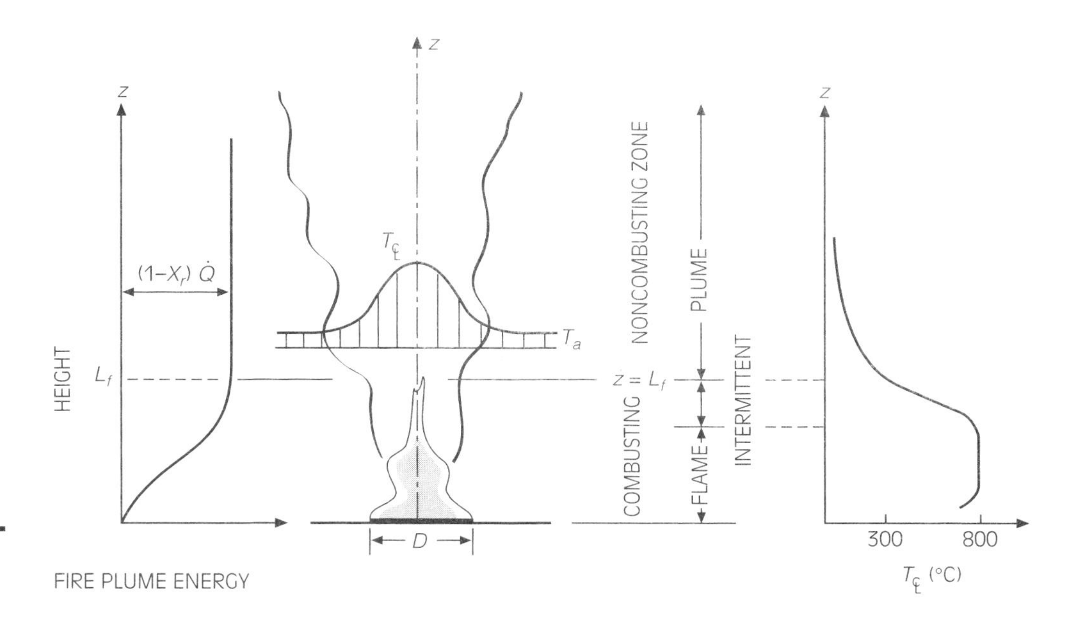

Figure 7-9 *Fire plume energetics.*

Consider the combustion zone where $z \le L_f$. The energy flow rate in the plume can be given in terms of the average plume temperature T at height

$$\dot{Q} = \dot{m}\, c_p\, (T - T_a),\ (\text{kW}) \tag{7-10}$$

where c_p is the specific heat of the plume gases, $\dfrac{(\text{kJ})}{(\text{gram} - {}^\circ\text{C})}$ and

$\dot{m}$ is the mass flow rate of the plume gases (gram/s).

The mass flow rate in the plume is nearly equal to the air entrainment rate. The energy released due to the combustion process is

$$\dot{Q} = \dot{m}_f\, \Delta H_c\, (1 - X_r),\ (\text{kW}) \tag{7-11}$$

where $\dot{m}_f$ is the mass rate of fuel reacted up to height z, (g/s) and

ΔH_c is the heat of combustion, $\dfrac{(\text{kJ})}{\text{gram}}$.

It has been stated that more air than stoichiometrically required is entrained over the flame. This is due to turbulence affecting the perfect mixing of air and fuel. We can therefore write that the mass air entrainment rate is

$$\dot{m} = n\, s\, \dot{m}_f \tag{7-12}$$

where s is the stoichiometric mass air to fuel ratio and n is the multiple of stoichiometric air required due to inefficient mixing. By equating Equations (7-10) and (7-11) and substituting Equation (7-12), it follows that

$$T - T_a = \frac{\Delta H_c (1 - X_r)}{c_p n s} \tag{7-13}$$

Since $\Delta H_c / s$ is approximately constant (3 kJ/g air) for most all hydrocarbon fuels and $c_p \approx 10^{-3}$ kJ/gK for air

$$T - T_a = \frac{3000\,(1 - X_r)}{n} \tag{7-14}$$

This result demonstrates that the temperature within the combusting zone is not dependent on fuel type except for variations in X_r. We saw that X_r can range from 0.15 to 0.60 for different fuels and may also depend on the scale of the pool fire (Tables 3-3 and 3-4). Large pool fires or very sooty fuels may have the tendency to shroud the luminous flame in cooler black soot as seen in Figure 7-4. This shroud of soot can block the outgoing flame radiation, making X_r smaller and T larger. This result has been found for large pool fires.

The entrainment rate of air and the inability of the fuel and air to efficiently mix is primarily a result of fluid mechanics—buoyancy and turbulence. Consequently, these processes should not strongly depend on the fuel burned. The quantity of extra stoichiometric air (n) is expected to remain constant for similar size pool fires. However, it has not accurately been measured, and experiments place it between 5 and 15. Such values would result in very low flame temperatures as computed from Equation (7-14), e.g., $n = 5$, $X_r = 0.2$ gives $T - T_a = 480°C$ or $T = 500°C$ for $T_a = 20°C$. Equation (7-14) represents an average temperature across the entire fire plume at height $z \leq L_f$.

The maximum temperature at the center line is higher, and values have been correlated in terms of the combustion length (see Equation (7-8)) and z as

$$\frac{T - T_a}{T_a} \sim \text{function } \frac{z}{(\dot{Q} / \rho_a c_{pa} T_a \sqrt{g})^{2/5}} \tag{7-15}$$

for a wide range of pool fire data.

Figure 7-10 represents this correlation shown for propane fires of 15 kW to 60 kW. Although the data are scattered to some degree, the line represents the correlation distinctly in three regions: combusting, intermittent, and noncombusting plume. The end of the intermittent zone—the maximum flame tip—is at $z/\dot{Q}^{2/5} = 0.20$ m/kW$^{2/5}$ which is consistent with the coefficient 0.23 of Equation (7-9). The correlation shows a maximum (time-averaged) turbulent flame temperature of approximately 800°C, and a centerline temperature of slightly over 300°C at the tip of the flame.

After the combustion zone, from Equations (7-10) and (7-11)

$$T - T_a = \frac{\dot{m}_f \Delta H_c (1 - X_r)}{\dot{m} c_p} \tag{7-16}$$

it can be seen that T falls rapidly as the air entrainment rate, $\dot{m}$, increases as more air is drawn into the plume over height z. In the noncombusting region, the fire plume energy is constant as indicated in Figure 7-9. The air entrainment rate controls the combustion and the decay in temperature after combustion.

Although Figure 7-10 only contains data from one fuel type, other experimental results for hydrocarbon fuels have shown the approximate universality of the graph. It is also consistent with the simple illustrative theory given by Equation (7-14). However, as the pool fire diameter becomes larger than 2 m to 4 m, as a benchmark, the sooting can reduce X_r causing maximum centerline turbulent flame temperatures as high as 1,200°C. Except for this higher plateau in the flame region of Figure 7-10, the rest of the correlation still applies. More important, for hydrocarbon fuels, the turbulent flame temperature is independent of fuel and can generally be taken as approximately 800°C for moderate fire sizes. Therefore, gasoline and wood fires should show no temperature differences, but their flames lengths may be significantly different. These flames have a similarity. They can be stretched and squeezed, but their temperature distributions are the same relative to the combusting and noncombusting zones.

■ NOTE

For hydrocarbon fuels, the turbulent flame temperature is independent of fuel and can generally be taken as approximately 800°C for moderate fire sizes.

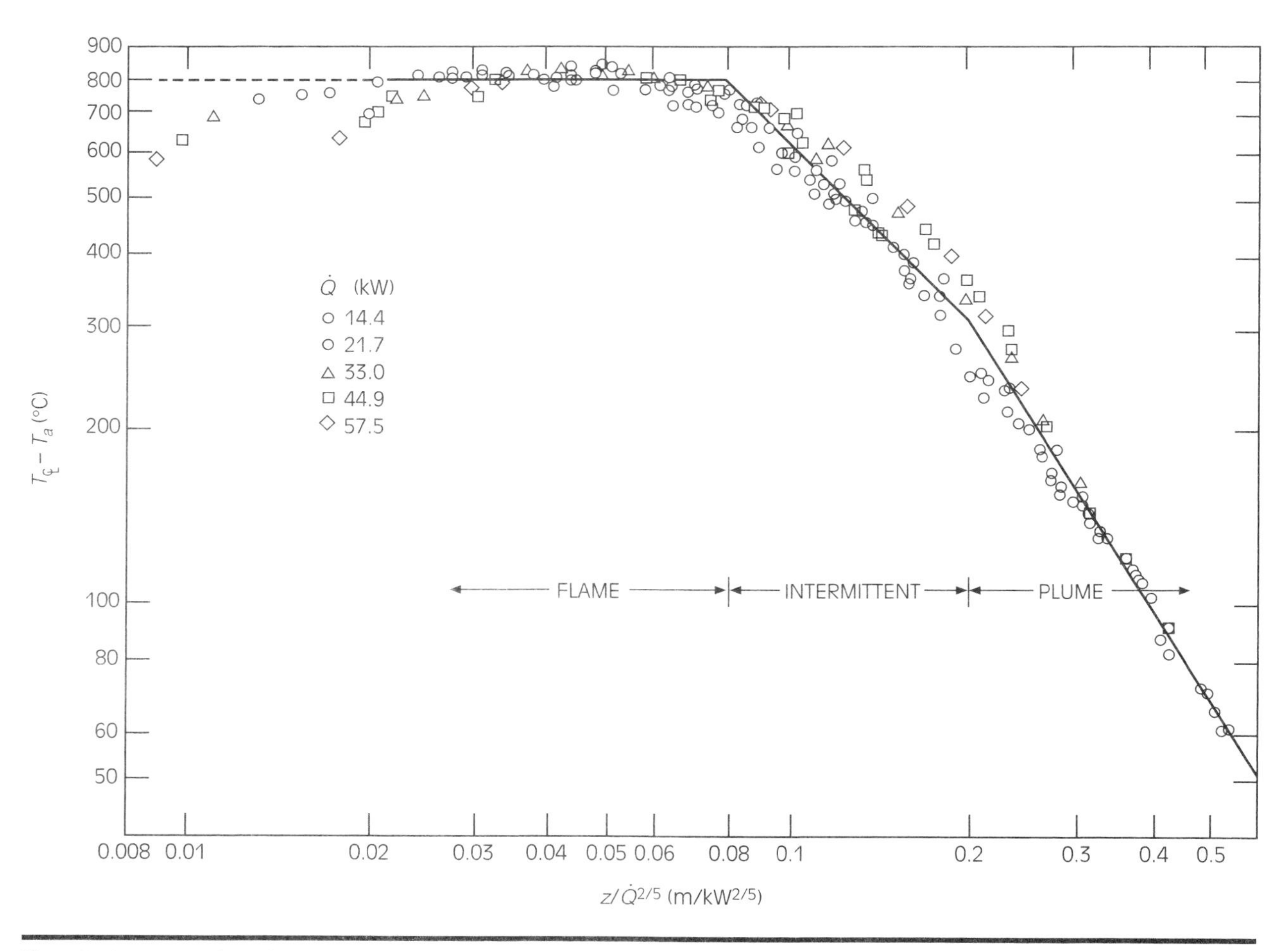

Figure 7-10 *Centerline fire plume temperatures. After McCaffrey, Ref. 1.*

FLAME HEIGHT AND TEMPERATURE CALCULATIONS

Compute the average flame height, and the centerline fire plume temperature at this flame height for the four fuels, 1 m in diameter. These fires were previously used in Chapter 6. The fuels are wood, polystyrene, heptane, and gasoline. See Figure 7-11 as an illustration of the example. Using Equation (7-9),

$$L_f = 0.23\ \dot{Q}^{2/5} - 1.02D$$

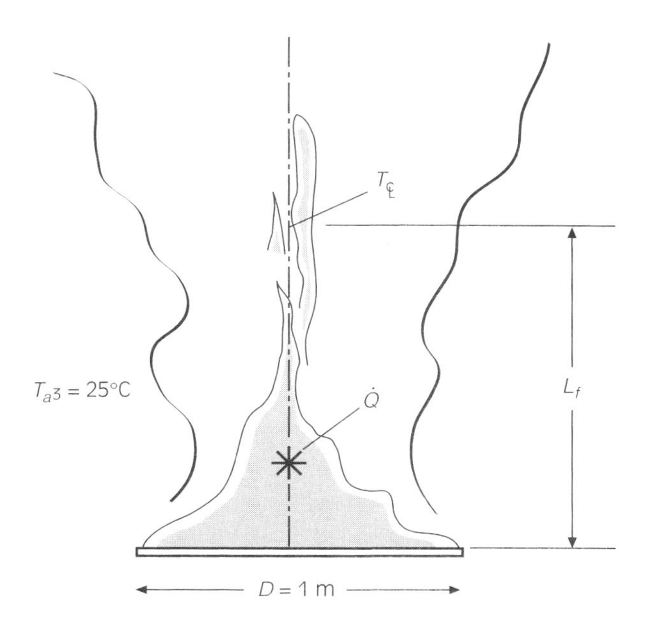

Figure 7-11 *Example to compute L_f and $T_{℄}$ at $z = L_f$.*

where $\dot{Q}$ is in kilowatts, and $D = 1$ m. Use the values of $\dot{Q}$ computed in Chapter 6. The flame height results follow:

Wood:

$$L_f = 0.23(130\ \text{kW})^{2/5} - 1.02(1\ \text{m}) = 0.59\ \text{m}$$

Polystyrene:

$$L_f = 0.23(1189\ \text{kW})^{2/5} - 1.02(1\ \text{m}) = 2.89\ \text{m}$$

Heptane:

$$L_f = 0.23(2661\ \text{kW})^{2/5} - 1.02(1\ \text{m}) = 4.37\ \text{m}$$

Gasoline:

$$L_f = 0.23(1887\ \text{kW})^{2/5} - 1.02(1\ \text{m}) = 3.68\ \text{m}$$

The temperature is found from Figure 7-10. Although flame radiation varies for the fuels, the graph has been found be approximately applicable to all fuels independent of X_r. This is not perfectly correct, but we do not have better results to work from at this time. The following fire plume temperatures at $z = L_f$ on the centerline have been computed for $X_r = 0$.

Fuel	T(°C)
Wood	745
Polystyrene	405
Heptane	355
Gasoline	375

The wood result is indicative of the continuous flame region, while the others are indicative of the peak flame height. There is probably some inconsistency in the wood crib case because these pool fire formulas do not explicitly account for the crib height.

The Nature of Turbulent Flame Temperature

adiabatic flame temperature
the maximum possible temperature in the reacting zone with no heat lost

We have just seen that the turbulent flame temperature for pool fire (and it could be shown for other fire configurations) is nearly a constant maximum of approximately 800°C and independent of fuel, particularly hydrocarbons. Literature values of **adiabatic flame temperatures** for hydrocarbon fuels burning in air typically range from 2,000°C to 2,300°C with measured laminar flame temperature of 1,800°C to 2,000°C. Why are turbulent fires 800°C, or 1,200°C at most? To understand this, we must understand what is being measured and therefore sensed by a thermometer or other heated surface.

The adiabatic flame temperature is the maximum possible temperature that can be achieved for that combustion process. Adiabatic means perfectly insulated; there is no heat loss from the reaction zone. As shown in Chapter 2, the candle flame is representative of a laminar diffusion flame. It is thin and reasonably steady in its position. Because it is thin, not much heat is lost by radiation. Primarily, heat is lost by conduction through the gas. Accordingly, we might expect its temperature to be slightly below the adiabatic flame temperature. Values of 1,800°C to 2,000°C are typical of such laminar flames. If by cooling or dilution, its temperature drops to about 1,300°C it is likely to extinguish. Then, how can a turbulent flame exist?

Figure 7-12 illustrates the measurement process for laminar and turbulent flames. A probe at a fixed point attempts to measure the flame temperature. The relatively stationary laminar flame is an easy target for the temperature probe. The fluctuating turbulent flame is not. The probe can be in and out of the flame on either the air or fuel side. A typical trace of the probe output over time is illustrated. Even a small probe may not respond fast enough in the fluctuating flame so the true maximum and minimum temperatures may not even be realized. The

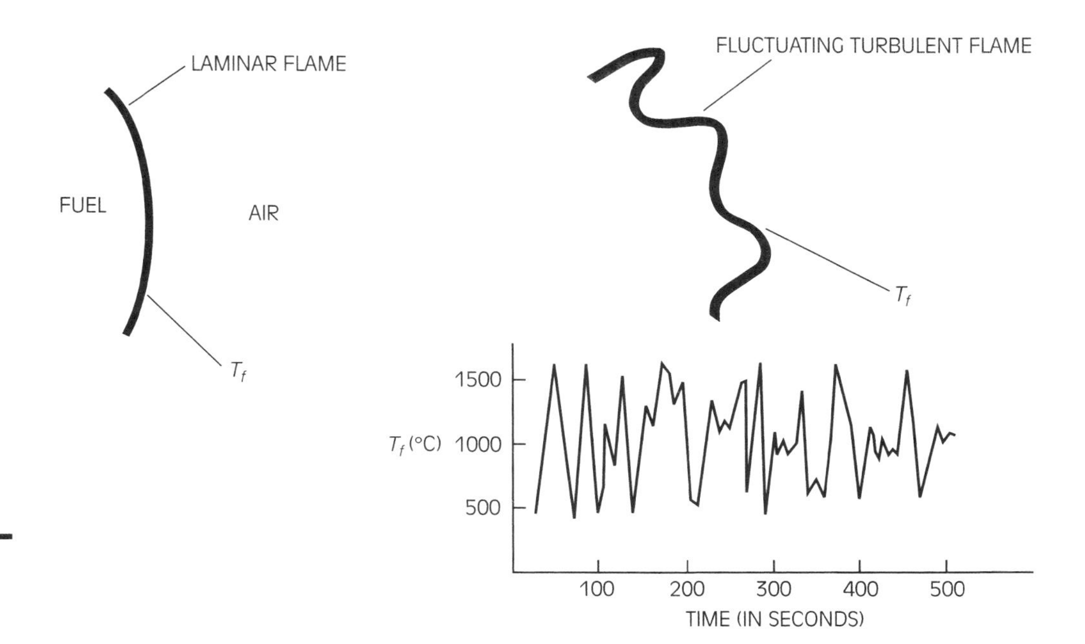

Figure 7-12
Measuring flame temperatures.

mean time or average temperature is what is reported in Figure 7-12, and what would be reported by a slowly responding temperature probe. The 800°C represents a time-average temperature. If we were capable of "riding" on the fluctuating flame, we would measure temperatures in excess of 1,300°C and closer to the laminar, flame temperatures. Note the thermal threat of this flame is represented by the time-average temperature. That is the temperature an object within or exposed to the turbulent flame would sense from the turbulent flame.

Summary

Buoyancy plays a significant role in natural fire plumes. Unlike jet flames, buoyant turbulent flame lengths depend on the energy release rate due to combustion, $\dot{Q}$. Flame length for turbulent pool fires is proportional to $\dot{Q}^{2/5}$, and plume temperatures depend on $z/\dot{Q}^{2/5}$ for each height z. Laminar and turbulent flame temperatures differ, and flame temperatures do not strongly depend on fuel type for hydrocarbon type fuels. Maximum flame temperatures in turbulent flames of moderate size do not exceed 800°C.

Because temperatures decrease very rapidly after combustion due to air entrainment, flame height is a critical measure of hazard. Objects above the flame are not likely to ignite; flame contact is generally required except for remote ignition by radiant heat of very large fires.

Review Questions

1. For a 500-kW flame compute the flame height if its base is 10 cm and its diameter is 100 cm.
2. A ceiling detector will alarm at a gas temperature of 150°C. For a fire directly under it, at 3-m distance, find the smallest energy release rate fire source that can cause the alarm.

True or False

1. Although a turbulent flame temperature is roughly 800°C, the temperature at the fluctuating flame tip is no more than 350°C.
2. Eddies or vortexes are rotating regions of flow.
3. Gasoline fires are much hotter than wood fires.
4. The velocity of gasoline vapors leaving the surface of a fire are launched at 3 m/s or more.
5. Theoretically, a buoyant plume can rise forever in a cold atmosphere.

Activities

1. Observe buoyant plumes in the atmosphere from smoke stacks, fires, or other sources. Note their nature, eddy characteristics, and how they are affected by wind.
2. In a fireplace, hold (with tongs) a piece of cardboard above the flames. See at what point of height in flames the cardboard can ignite.

3. In a laboratory setting under a hood, ignite a small pan of heptane or similar flashpoint liquid fuel. Use a pan no larger than 10 inches in diameter. Observe the fluctuations of the flame, its necking in at its base, and the eddy vortices.

References

1. B. J. McCaffrey, *Purely Buoyant Diffusion Flames: Some Experimental Results*, NBSIR 79-1910 (Gaithersburg, MD: National Bureau of Standards, 1979).
2. H. C. Hottel and W. R. Hawthorne, "Diffusion in Laminar Flame Jets," in *Third Symposium on Combustion and Flames and Explosion Phenomena*, (Baltimore, MD: Williams & Wilkins, 1949), 254–266.
3. B. J. McCaffrey, "Flame Height," chap. 2-1 in *SFPE Handbook of Fire Protection Engineering*, 2d ed., edited by P. J. DiNenno (Quincy, MA: National Fire Protection Association, June, 1995).
4. G. Heskestad, "Luminous Heights of Turbulent Diffusion Flames," *Fire Safety Journal* 5 (1983): 103–108.

Combustion Products

Learning Objectives

Upon completion of this chapter, you should be able to:

- Understand the nature and levels of combustion products for various fire conditions.
- Explain the property called *yield* and how it relates to concentration of that product in smoke.
- Quantitatively understand the hazards of combustion products in smoke.

INTRODUCTION

products
chemical compounds produced by fire

yield
the mass of product produced per unit mass of fuel supplied

species
another name for distinct chemical compounds and other molecular structures in a mixture, usually gases

concentration
the percentage of material per unit mass (or volume) of its mixture

We have seen that the mass burning rate is a significant quantity that results from the burning rate flux ($\dot{m}''$) and the area (A) involved in burning due to ignition and spread: $\dot{m} = \dot{m}'' A$. From this burning rate ($\dot{m}$) we can derive all the **products** that result from the fire's chemical reaction. We have seen that energy release rate ($\dot{Q}$) is derived as $\dot{m}\Delta H_c$ where ΔH_c is the heat of combustion. $\dot{Q}$ controls the smoke temperature and the surrounding heat flux. It causes an immediate effect on its surroundings. All of the other products of the fire's chemical reaction are frozen in the smoke emanated in the buoyant plume leaving the fire. As with $\dot{Q}$ proportional to $\dot{m}$ by ΔH_c, the rates of production of the other products are proportional to $\dot{m}$ according to the fire's chemical reaction. This proportionality is called the **yield** for each product **species**. These products of combustion involve carbon dioxide and water vapor for the complete burning of hydrocarbon fuels, however, other products are possible due to incomplete combustion or due to the presence of other elements in more complex fuels. These combustion products can have a variety of harmful effects on people, on commodities, and on equipment. Usually the harm or damage is related to the amount (or **concentration**) of combustion product and the time of exposure. In this chapter, we present quantitative criteria for several products of combustion, assuming that the student has an elementary background in chemistry.

SCOPE OF COMBUSTION PRODUCTS

completeness
pertaining to a combustion process going to its most stable state, an ideal reaction

chemical equation
an equation showing the fuel plus oxygen-producing products

The nature of the combustion products depends on the fuel composition and on the fire process. We have seen that fire can be characterized as (1) a premixed flame, (2) a diffusion flame, (3) smoldering or surface oxidation, and (4) spontaneous ignition leading to either (2) or (3). Only flaming and smoldering are considered here. Ultimately, we need to represent the chemical reaction occurring in the fire process. This reaction depends on the **completeness** of the reaction. For example, if we burned methane in air (recall 1 mole of air is 0.21 moles of O_2 and 0.79 moles of N_2), we can write a **chemical equation**:

$$CH_4 + \frac{(4)}{0.42}(0.21\ O_2 + 0.79\ N_2) \rightarrow CO_2 + 2H_2O + \frac{(4)(0.79)}{0.42} N_2$$

For complete or ideal combustion, the products, by necessity and definition, are CO_2 (carbon dioxide) and H_2O (water). At combustion temperatures, these are gases, although the water could condense if it finds a cool surface. The coefficients of each term in the chemical equation have been deduced so that atoms (e.g., 1 C atom and 4 H atoms of CH_4) are not destroyed. From this balance we can compute for every 16 g of CH_4 (recall 1 mole of CH_4 has a mass of 16 g, 12 g for C and 1 g for each H) that we must have

$$\frac{(4)}{0.42}[0.21(32)+0.79(28)]$$

or 274.7 g of air. (Note 1 g of air has 0.23 g of O_2 and 0.77 g of N_2.) These proportions insure all of the methane and oxygen are completely consumed in the reaction. The ratio of this mass of air to mass of fuel is called the (ideal) stoichiometric air to fuel mass ratio (s), i.e.,

$$s=\frac{274.7 \text{ g air}}{16 \text{ g } CH_4}=17.2$$

overventilated more than stoichiometric air is available

underventilated less than stoichiometric air is available

If more air is available, the condition is said to be **overventilated** or the process is fuel lean. For less air, the process is termed **underventilated** or fuel rich. Notice that in all cases the nitrogen in the air is passive and remains constant. However, it will be heated as a result of the chemical reaction's energy release.

Typical fuels involved in fire processes are hydrocarbons, that is, composed of hydrogen (H) and carbon (C). Other substances may also include oxygen (O), nitrogen (N), chlorine (Cl), fluorine (F), and bromine (Br). The resultant combustion products depend on the process as illustrated in Figure 8-1. In an ideal complete reaction, the N, Cl, F, and Br would yield their respective gases. However, in any fire process the reaction is not complete, and the resultant gases normally

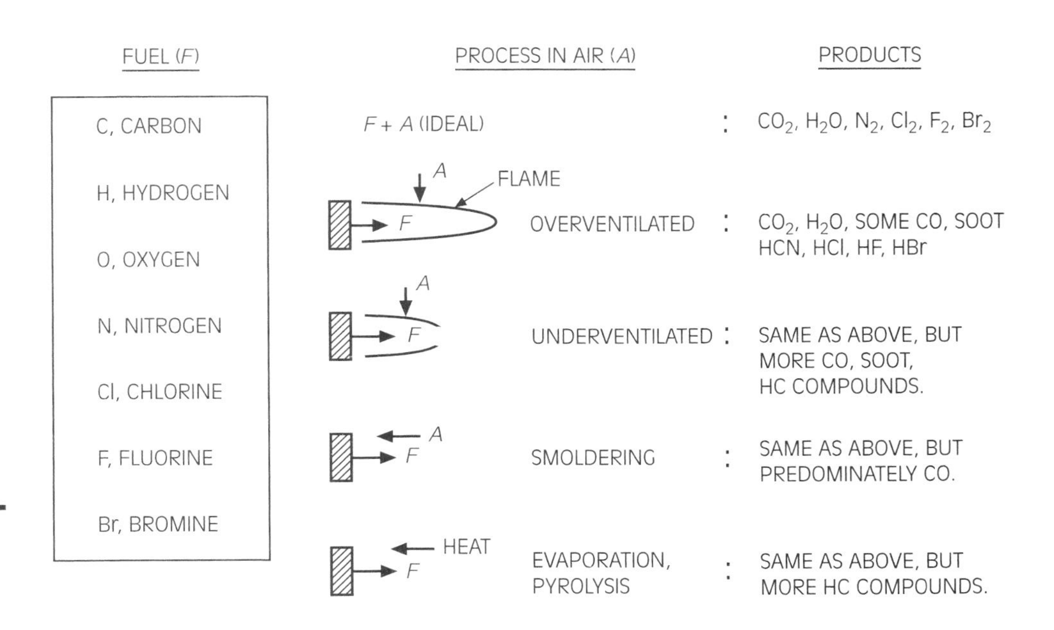

Figure 8-1 *Typical products of combustion in fire.*

■ NOTE
Smoldering is a process that requires very little air to maintain.

are hydrogen cyanide (HCN), hydrogen chloride (HCl), hydrogen fluorine (HF), and hydrogen bromine (Br), respectively. In addition, incomplete combustion leads to carbon monoxide (CO) in place of carbon dioxide (CO_2), soot (principally carbon), and many hydrocarbons (HCs) as a result of the thermal decomposition of the original fuel. Figure 8-1 shows the various fire processes and the associated products of combustion. It is very important to recognize the fire process because the products of combustion depend on the process as well as on the fuel. Smoldering requires very little air to maintain. Its combustion is incomplete resulting in a significant yield of CO.

Initially there is sufficient air available in a room fire to support flaming combustion. If fire occurs in a closed room, the oxygen in the air can be expended and the process becomes underventilated. Extinction will occur if no additional air becomes available. A small opening to the room will bring in some air, but if the fuel generated is too great the fire will remain underventilated.

equivalence ratio, Φ
the ratio of fuel to air times the stoichiometric ratio, s; or (fuel/air) available divided by (fuel/air) stoichiometric

The parameter that quantitatively represents the over- and underventilation states is called the **equivalence ratio**, Φ, defined as the mass ratio of fuel (gas) available to air available times the stoichiometric ratio, s.

$$\Phi = \left(\frac{\text{mass of fuel (gas)}}{\text{mass of air}} \right) \times s \tag{8-1}$$

That is, if $\Phi > 1$, the fire is underventilated. As we saw for the candle flame that was quenched by the metal screen, the flame causes thermal decomposition of the fuel before the primary combustion process occurs in the flame. For $\Phi > 1$, these decomposition products are not burned to completion. For $\Phi < 1$, there is sufficient air and the process will be complete: This case is said to be overventilated. All fires start with $\Phi < 1$ and with continued growth can reach $\Phi > 1$ states. The chemistry of these processes is complex, and consequently, empirical data must be used to quantify the nature of the combustion products for each fuel.

YIELDS

As the heat of combustion gives us the energy release per unit mass of fuel burned, yields give us the mass production of each product specie per unit mass of fuel burned. For example, the yield of CO is defined as

$$y_{CO} = \frac{m_{CO}}{m} \tag{8-2}$$

where m_{CO} is the mass of CO produced and m is the mass of fuel burned. Alternatively, we could consider the rates of mass, i.e., $y_{CO} = \dot{m}_{CO}/\dot{m}$, where $\dot{m}$ is the rate of burning and $\dot{m}_{CO}$ is the rate of CO mass produced. It is generally accepted that the yields are fairly constant for a given fuel as long as $\Phi < 1$. But as the fire condition becomes underventilated the yields change.

Tewarson[1] of Factory Mutual Research Corporation has nearly single-handedly established a wealth of consistent data on the yields of various fuels. Although mainly derived from relatively small-scale tests but over a range of oxygen and heating conditions, it is expected that these results apply to realistic fire conditions. Table 8-1 lists a sample of his data for representative fuel classes: gases, liquids, and solids. The yields are shown for CO_2, CO, and soot along with the measured heat of combustion, ΔH_c. These are experimentally derived values representative of the fire process. The ΔH_c is less than its theoretical value that would occur for the ideal complete reaction indicated in Figure 8-1. These ideal values are found in the literature and do not necessarily represent the energy released for the particular fire condition. For comparison, the ideal values of ΔH_c are also listed in Table 8-1.

In most cases, the yields (or negative yields, as in the consumption of oxygen) are slightly less than their ideal chemical reaction values for CO_2, H_2O, and

Table 8-1 *Fuel properties as a function of ventilation.*

	Overventilated Conditions						Underventilated Conditions		
Fuel	y_{CO_2} (g/g)	y_{CO} (g/g)	y_{soot} (g/g)	ΔH_c (kJ/g)	$\Delta H_{c,\,ideal}$ (kJ/g)	D_m (m²/g)	y_{CO} (g/g)	y_{H_2} (g/g)	y_{HCl} (g/g)
Gases:									
Propane	2.85	0.005	0.024	43.7	46.4	0.155	0.229	0.011	—
Acetylene	2.6	0.042	0.096	36.7	48.2	0.315	NA*	NA	—
Liquids:									
Ethyl alcohol	1.77	0.001	0.008	25.6	26.8	NA	0.219	0.0098	—
Heptane	2.85	0.010	0.037	41.2	44.6	0.190	NA	NA	—
Solids:									
Wood (Red oak, pine)	1.27	0.004	0.015	12.4	17.7	0.037	0.138	0.0024	—
Polymethyl methacrylate (PMMA)	2.12	0.010	0.022	24.2	25.2	0.109	0.189	0.0032	—
Polystyrene (PS)	2.33	0.060	0.164	27.0	39.2	0.335	NA	NA	—
Nylon	2.06	0.038	0.075	27.1	30.8	0.230	NA	NA	—
Polyurethane (PU) -flexible foam	1.51	0.031	0.227	19.0	27.2	0.326	NA	NA	—
Polyvinyl chloride (PVC)	0.46	0.063	0.172	5.7	16.4	0.400	0.36	NA	0.400

Source: Based on data from Tewarson, Ref. 1.

*NA = not available

incomplete combustion the chemical reaction stops before using up all of the species that can react with O_2

O_2. The yields associated with the products of **incomplete combustion** (e.g., CO, soot) are relatively small. But as we shall see, these are the culprits that do the damage.

In addition the property D_m, which pertains to visibility through smoke, is also listed in Table 8-1. We return to this quantity later in the chapter.

ventilation limited underventilated

Some typical underventilated values of yields for CO and H_2 are listed in Table 8-1, but their precise values depend on Φ. For $\Phi < 1$, we have the Table 8-1 overventilated values. Figure 8-2 displays how the underventilated yields vary with Φ. These data also come from Tewarson[1] and are very valuable. For example, they can explain how carbon monoxide is increased as the fire becomes underventilated (a state sometimes called **ventilation limited**). Also, the results show that CO yields depend on fuel, but the yields (+ or –) of O_2, soot, and CO_2 are essentially fuel independent when plotted as $\frac{y_{\text{under}}}{y_{\text{over}}}$ versus Φ.

Let us consider an example. In chapter 6 we computed the burning flux of wood and polystyrene (ps) as 11 and 38 g/m²s. Suppose we have the same size fires: 2000 kW. Since (from Equation 6.2)

$$\dot{Q} = \dot{m}'' A \, \Delta H_c$$

the fuel areas involved can be computed as

$$A = \frac{\dot{Q}}{\dot{m}'' \Delta H_c}$$

For wood

$$A = \frac{2000 \text{ kW}}{(11 \text{ g/m}^2\text{s})(12 \text{ kJ/g})} = 15.2 \text{ m}^2,$$

and for PS

$$A = \frac{2000 \text{ kW}}{(38 \text{ g/m}^2\text{s})(27 \text{ kJ/g})} = 1.9 \text{ m}^2$$

free-burning burning in open-air

For a fire in the open air (**free-burning**) or in a large room with sufficient open windows, the overventilation yields apply. The production rate of CO is then computed by

$$\dot{m}_{\text{CO}} = y_{\text{CO}} \, \dot{m}'' A, \tag{8-3}$$

or for wood:

$$\dot{m}_{\text{CO}} = (0.004)(11 \text{ g/m}^2\text{s})(15.2 \text{ m}^2) = 0.67 \text{ g/s}$$

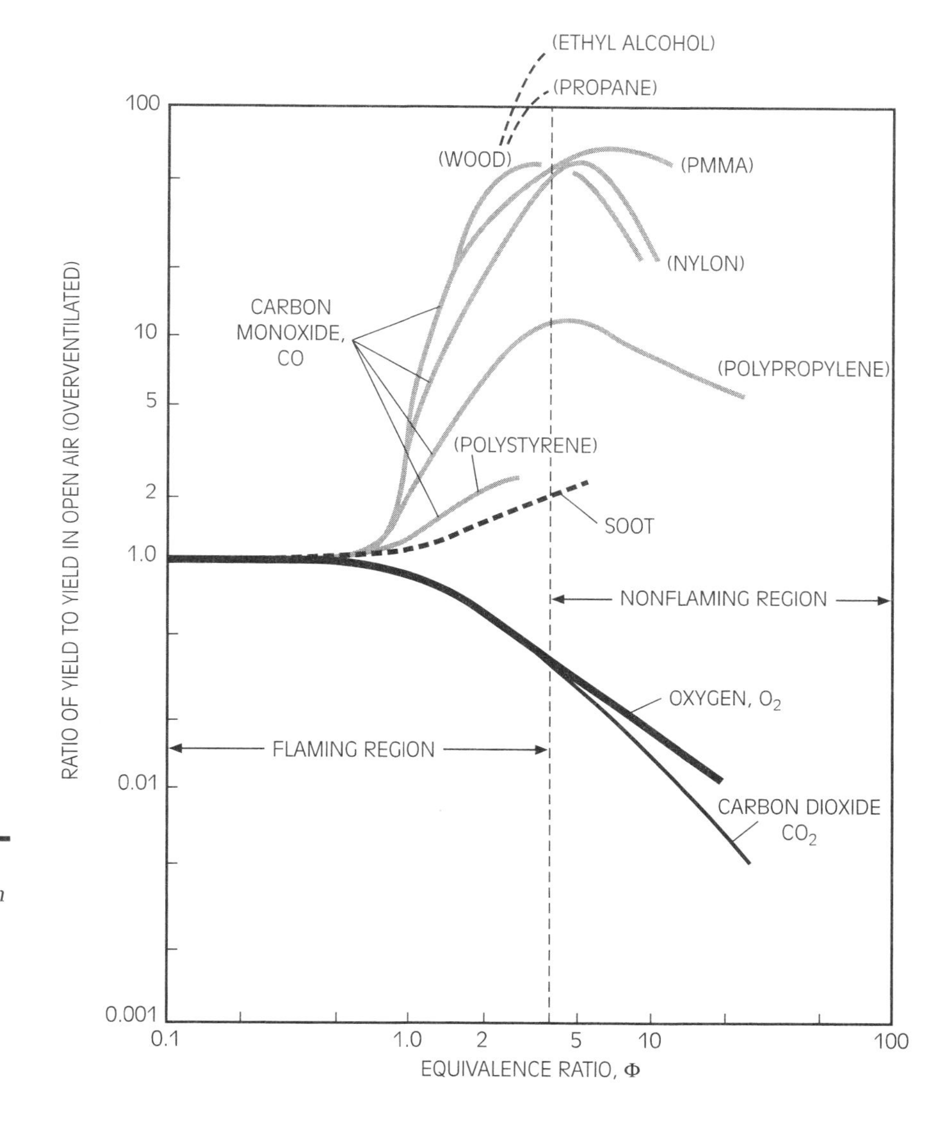

Figure 8-2 *Effect of underventilation on yields for many materials—approximate representation. After Tewarson, Ref. 1.*

and for PS:

$$\dot{m}_{CO} = (0.060)(38\ g/m^2S)(1.9m^2) = 4.3\ g/s$$

As a result, we see more CO is produced by PS.

But, what if these fires are in a room where the potential air supply through windows is 500 g/s, typical of a small broken window? Let us calculate Φ to examine the degree of underventilation. For this computation, we need s. Let us estimate s from the ideal reaction using $\Delta H_{c,\text{ideal}}$, which is derived as follows:

$$s \approx \frac{\Delta H_{c,\text{ideal}}}{3\ kJ/g_{air}} \tag{8-4}$$

using the approximation of a constant energy release for burning the oxygen in air. For most ordinary combustibles, approximately 3 kJ is released for each gram of air used by the fire reaction. Then, for wood:

$$s = \frac{17.7}{3} = 5.9\ g\ air/g\ wood$$

and for PS:

$$s = \frac{39.2}{3} = 13.1\ g\ air/g\ PS$$

The corresponding Φ values are

$$\Phi_{wood} = \frac{(11\ g/m^2s)(15.2\ m^2)(5.9)}{(500\ g/s)} = 1.97$$

and

$$\Phi_{PS} = \frac{(38\ g/m^2)(1.9\ m^2)(13.1)}{(500\ g/s)} = 1.9$$

From Figure 8.2 and the over-ventilated values from Table 8-1, we can estimate that for wood at $\Phi = 1.97$:

$$y_{CO} \approx 50\ (0.004) \approx 0.20$$

and for PS at $\Phi = 1.9$:

$$y_{CO} \approx 2.0\ (0.060) \approx 0.12$$

■ NOTE
CO production depends on the fuel as well as the air supply.

■ NOTE
Smoldering, especially of a cotton or polyurethane foam mattress or sofa, produces CO yields of as high as 0.2 to 0.3.

The amount of CO produced has doubled for the PS, but only increased by 50 times for the wood for this particular room fire condition. This example should illustrate that CO production depends on the fuel as well as the air supply. For the wood CO yield of 0.2, this state would give a CO production rate of 33.4 g/s provided the burning rate of the wood can be maintained at 11 g/m^{2}s at this underventilated state. This is likely, as long as the fire is maintaining a high heating condition on the wood.

Finally, it should be pointed out that smoldering, especially the notorious fire scenario of a cotton or polyurethane foam mattress or sofa, produces CO yields of as high as 0.2 to 0.3.

CONCENTRATIONS

smoke
gases, no longer chemically reacting, that emanate from the fire

The yields measure what is produced at the fire source. The principal hazard of fire is the composition and associated concentrations of the **smoke**. By smoke we mean the gas stream that flows away from the fire (as illustrated in Figure 8-3) continuing to mix with air with no further chemical reactions. The amount of mixing with air determines the concentration. Associated with each species yield we have a corresponding concentration. Moreover, associated with the energy yield, ΔH_c, we have the concentration-like variable, temperature. Like some species, temperature decays as well due to losses (of heat) as the smoke encounters clean, cold surfaces. For example, soot is deposited, and HCl (gas) can condense on surfaces. This surface transfer is increased as the local smoke velocity increases; that is why heavy depositions of soot are seen around door cracks as the smoke squirts through at high velocity.

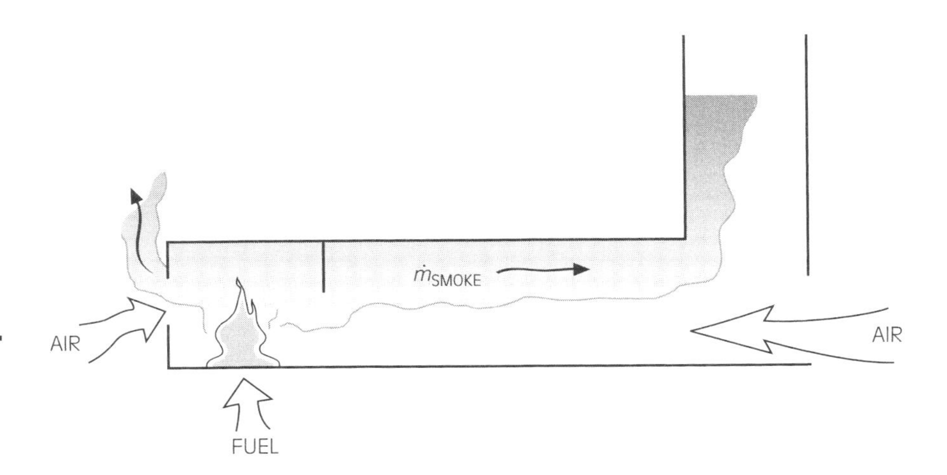

Figure 8-3
Illustration of smoke movement from a fire.

mass concentration, *Y* ratio of species mass to mixture mass; also referred to as mass fraction

volume fraction, *X* species concentration based on volume

Concentration can be expressed in several ways. **Mass concentration** is defined as

$$Y_{\text{species}} = \frac{\dot{m}_{\text{species}}}{\dot{m}_{\text{smoke}}} = \frac{Y_{\text{species}}\dot{m}}{\dot{m}_{\text{smoke}}} \tag{8-5}$$

where $\dot{m}_{\text{smoke}}$ is the mass flow rate of the smoke. Usually concentrations are represented as **volume fractions**, i.e., the volume occupied by the species at normal atmospheric temperature and pressure to the corresponding smoke volume. For example, if we could extract an individual species and let it expand to normal atmospheric conditions, then its volume compared to the volume resulting from extracting the entire smoke mixture is volume fraction. This volume fraction can be estimated as

$$X_{\text{species}} \approx \frac{29\,Y_{\text{species}}}{M_{\text{species}}} \tag{8-6}$$

where M_{species} is the molecular weight of the species and 29 is an approximate molecular weight of the smoke (since it mostly contains N_2, 28 g/mole).

In the previous example for production of CO due to the PS fire, the flow rate of the smoke flow must balance the fuel and air flow rate, so that $\dot{m}_{\text{smoke}} = 500 + (38\ \text{g/m}^2\text{s})(1.9\ \text{m}^2) \approx 572\ \text{g/s}$.

Therefore, from Equations (8-3) and (8-5),

$$Y_{\text{CO, over}} = \frac{4.3\ \text{g CO/s}}{572\ \text{g/s}} = 0.0075 \text{ or } 0.75\%$$

for the overventilated case and

$$Y_{\text{CO, under}} = \frac{(0.12)(38\ \text{g/m}^2\text{s})(1.9\ \text{m}^2)}{572\ \text{g/s}} = 0.015 \text{ or } 1.5\%$$

for the underventilated case. The corresponding volume concentrations are

$$X_{\text{CO, over}} = \frac{(0.75\%)(29)}{(12+16)} = 0.78\%$$

and

$$X_{\text{CO, under}} = \frac{(1.5\%)(29)}{(28)} = 1.56\%$$

Let us examine the hazard these concentrations present. It is common to

parts per million, ppm
concentration based on 10^6 parts of the mixture

express low concentrations (by volume) in **parts per million** (ppm). For example, the CO volume fraction given above as 0.0075 is 7500 ppm because $0.0075 = \frac{7500}{10^6}$.

HAZARDS

Each concentration of species can present a specific hazard. The hazard can be to humans, to equipment, and to property. Here we consider only the hazard to humans. However, essentially in each case, the hazard is measured by the duration of the concentration exposed. Below a threshold concentration, nothing is damaged. Here we use concentration in a general sense. It can include temperature, the reduction in oxygen (**vitiation**, $0.21 - X_{O_2}$), as well as specific harmful combustion product concentrations such as carbon monoxide, X_{CO}.

vitiation
refers to the reduction of oxygen concentration in air

Combustion products such as soot and odor-bearing hydrocarbons can destroy significant quantities of goods, even when the actual fire damage by heat is small. Soot ladened with HCl or other acid gases (e.g., HBr, HF) can deposit on equipment. If not cleaned, this acidity can cause considerable damage later through corrosion. Such effects on goods and equipment we do not pursue further.

We draw from the review by Purser[2] to examine toxicity effects.

narcosis
effect of inducing sleep

Narcotic Gases

Narcosis is the state of induced sleep, which reduces the capability of escape and can lead to death. It can be caused by the combustion gases CO, HCN, and even CO_2 at high concentrations. It can also be induced by the reduction of oxygen. All animals require oxygen to be transported in the body by **oxyhemoglobin** (O_2Hb). The introduction of CO into the lungs causes the **hemoglobin** to attach instead to the CO as **carboxyhemoglobin** (COHb), which prevents the oxygen from attaching. The introduction of HCN affects the cells, and also prevents oxygen from being used by the body. Roughly, all of these effects are additive, so that if you are half dead from CO, 3/8 dead from HCN, and 1/8 dead from the reduction of oxygen, you are dead.

oxyhemoglobin
oxygen-bearing hemoglobin

hemoglobin
compound in blood that transports oxygen or carbon monoxide

carboxyhemoglobin
carbon-monoxide-bearing hemoglobin

Figure 8-4 shows the time to reach loss of consciousness for exposure to HCN and CO in primates for a specific activity level. We see from this figure that HCN is more toxic: It requires much less concentration to cause unconsciousness. Also the threshold levels below which no loss of consciousness occurs is about 90 ppm for HCN and 900 ppm (or 0.09%) for CO. Below these levels, incapacitation is still possible. Table 8-2 shows the effect of oxygen loss due to oxygen or carbon monoxide on humans. Incapacitation can occur at as low as 8–12% O_2 and 0.015–0.040% CO corresponding to 20–40% COHb. Usually 40% COHb indicates an incapacitation level and 60% COHb a lethal state. These levels are not precise and can vary in humans depending on health and other factors.

■ NOTE
HCN is more toxic than CO.

■ NOTE
Carbon monoxide is the leading killer of people in fire.

Carbon monoxide is the leading killer of people in fire, principally because

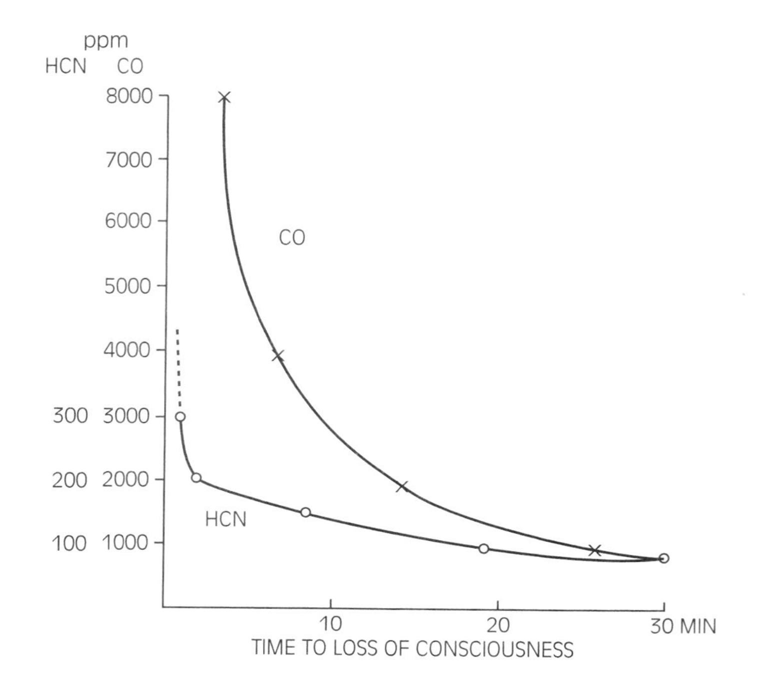

Figure 8-4 *Time to unconsciousness in primates by exposures to CO or HCN. After Purser, Ref. 2.*

Table 8-2 *Effects of O_2 loss in blood.*

Due to Decrease in Oxygen Concentration		Due to Increase in Carbon Monoxide Concentration		
O_2Hb (%)	**X_{O_2} (%)**	**COHb (%)**	**X_{CO} (%)**	**Effect**
90–100	15–21	0–10	<0.008	None
80–90	12–15	10–20	0.008–0.015	Fatigue
60–80	8–12	20–40	0.015–0.04	Dizziness, nausea, possible paralysis
50–60	6–8	40–50	0.04–0.06	Prostration, asphyxiation, collapse
30–50	3–6	50–70	0.06–0.3	Unconscious in minutes, possible death
0–30	0–3	70–100	>0.3	Unconscious in seconds, death likely

Table 8-3 *Respiration minute volume (RMV) (liters/min.) for a 70-kg man.*

RMV	Activity
8.5	Resting
25	Light work
50	Heavy work, slow running

Source: Based on data from Purser, Ref. 2.

of the long-time exposure from smoldering fires (usually > 1 hour) or because of rapid fire growth resulting in an underventilated fire. An approximate formula that shows the effect of exposure to CO over time is given in terms of the resultant COHb:

$$\text{COHb}(\%) \approx 0.33\, RMV \cdot X_{CO}\,(\%) \cdot t(\text{min}) \quad \textbf{(8-7)}$$

where t is the exposure time in minutes,
X_{CO} is the volumetric CO concentration in %, and
RMV is the **respiration minute volume** or inhalation rate in liters/min.

resperation minute volume, RMV
inhalation rate

dose
the accumulation of product concentration over time

NOTE
Dose is usually a direct measure of harm for concentrations above threshold levels.

Table 8-3 gives typical RMV values. The significance of Equation (8-7) is that the damage property (COHb) depends on the exposure quantity ($X_{CO}t$) and the respiration rate (RMV). Not only does activity effect RMV, but so does the exposure to CO_2, a plentiful combustion product. Table 8-4 shows the effects of CO_2, both on RMV and on narcosis. The quantity Xt is commonly referred to as the **dose** of the product. Dose is usually a direct measure of harm for concentrations above threshold levels.

We see in a fire smoke environment with a hydrocarbon fuel, that the atmosphere can contain CO_2, reduced O_2, and CO. All three accumulate to induce narcosis due to reduced oxygen in the body, hypoxia. If N is contained in the fuel, HCN will be produced and it too adds to the narcosis. From measurements, CO is the predominant cause of death, but HCN can be a significant contributor. By the time CO_2 and O_2 become hazardous, the CO and HCN effects have done their damage.

irritant gases
acid gases and other hydrocarbon by-products that can cause pain to the eyes or nasal tract on inhalation

Irritant Gases

Irritant gases consist of the acid gases HCl, HF, HBr, and other hydrocarbon by-products, such as acrolein and formaldehyde which can be derived from wood. These gases can be painful to the eyes and air passages, but they are not likely to be fatal during inhalation. Postexposure fatality is more likely. Pulmonary dam-

Table 8-4 *Effects of CO_2 on humans at normal atmospheric pressure and O_2 concentration.*

CO_2 Concentration in Inhaled Air (%)	Effect
0.04	Normal air
0.5	Safe limit, prolonged exposure
1.8–2.0	30–50% increase in ventilation rate
2.5–3.0	100% increase in ventilation rate
4.0	300% increase in ventilation rate
5.0	Dizziness, poisoning symptoms, > 30 minutes
7.0–9.0	Unconscious, in 15 minutes
10.0–30.0	Unconscious, in < 10 minutes, followed by death

Source: From Purser, Ref. 2.

age can result in death up to a day following exposure. It is also possible that these gases tend to reduce the RMV.

Smoke Visibility

smoke visibility, L_v ability to perceive objects through smoke over a specific distance, L_v

Smoke visibility is perhaps the first effect in a fire, even before the fire becomes a serious thermal threat. Light is attenuated by smoke due to particles, mainly soot, but also due to tarry condensibles. Soot gives smoke a black color. Tars, generated by heating in an ample air supply, give smoke its whitish color. The property that measures the attenuation (absorption and scattering) in smoke is the extinction coefficient, κ_s. κ_s is measured in m^{-1}. It can be used to compute the reduction in light intensity, I, over a light path length, l. The original intensity, I_O is reduced by the smoke to

$$I = I_O \exp(-\kappa_s l) \quad \textbf{(8-8)}$$

The ability to see through this smoke is called visibility, L_v, and is related to κ_s. Figure 8-5 shows the relationship between these variables.

mass optical density, D_m optical property related to the yield of particulates in smoke

L_v is the maximum distance an object can be seen and recognized through the smoke. The quantity **mass optical density** (D_m) is directly related to the yield of solid and liquid particulates in the smoke. This property is easily measured and has been tabulated in Table 8-1. The extinction coefficient depends on the concentration of the smoke particulates and can be computed for a given burning rate ($\dot{m}$) and smoke volumetric flow rate $\dot{V}$.

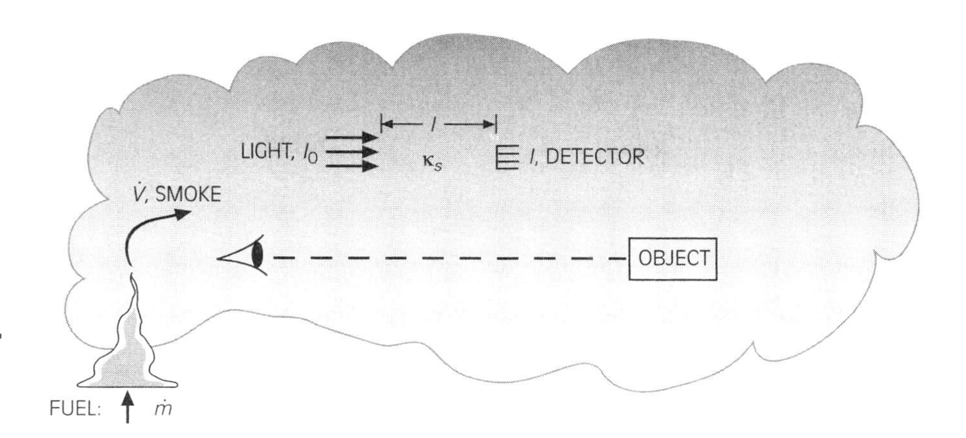

Figure 8-5 *Smoke attenuation and visibility.*

$$\kappa_s = \frac{\dot{m} D_m}{\dot{V}} \tag{8-9a}$$

for a flowing case, or

$$\kappa_s = \frac{m D_m}{V} \tag{8-9b}$$

for a static case where m is the mass of fuel burned in a closed volume, V.

Let us consider an extension to our previous example for the PS in the underventilated room fire. Assume the smoke has cooled to normal air temperatures so that the density of the cool gas mixture is approximately 1.0 kg/m³. Therefore,

$$\dot{V} = \frac{\dot{m}_{\text{smoke}}}{\rho} = \frac{572 \text{ g/s}}{1 \text{ kg/m}^3 \times 10^3 \text{ g/kg}} = 0.572 \text{ m}^3\text{/s}$$

From Table 8-1, $D_m = 0.34$ m²/g for the overventilated case. Equation (8-9a) gives

$$\kappa_s = \frac{(38 \text{ g/m}^2\text{s})(1.9 \text{ m}^2)(0.34 \text{ m}^2\text{/g})}{0.572 \text{ m}^3\text{/s}}$$

$$\kappa_s = 42.9 \text{ m}^{-1}$$

To consider the underventilated case at $\Phi = 1.9$. we could assume D_m is proportional to y_s. Then from Figure 8-2, D_m is increased by about 1.5 as is y_s.

For each value of κ_s, there is a corresponding visibility, which is given in Figure 8-6 as approximately

$$L_v = C_v / D_o$$

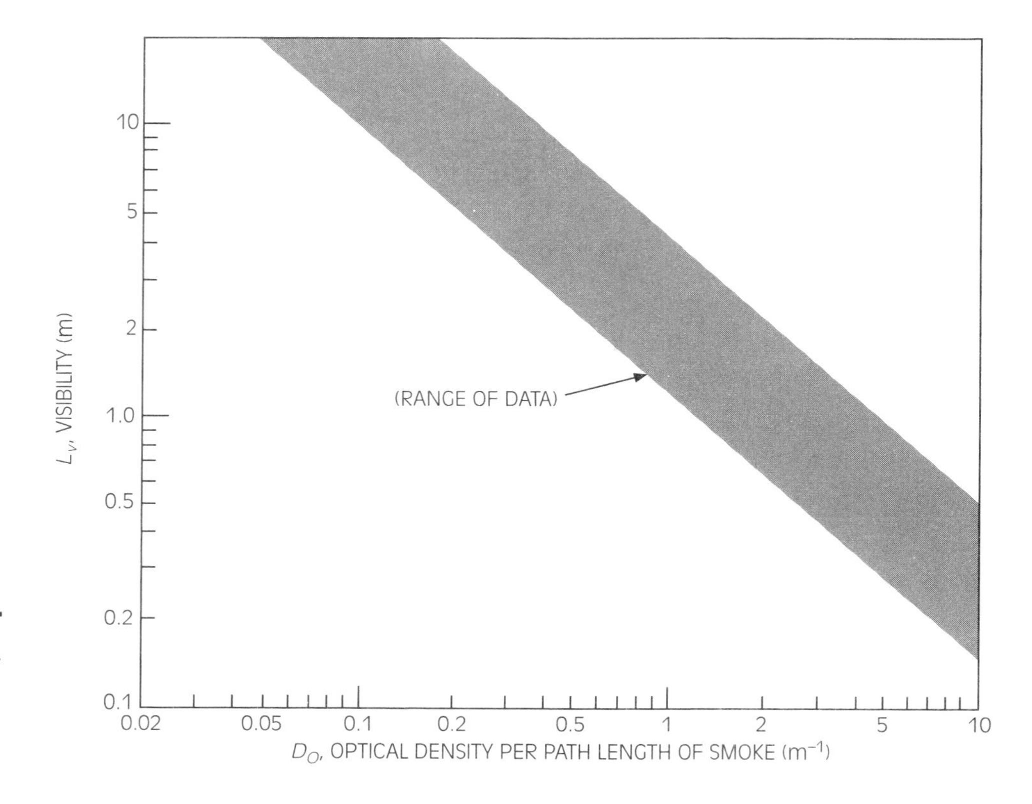

Figure 8-6 *Smoke visibility in terms of optical density per path length. After Quintiere, Ref. 3.*

where D_o is the optical density per unit path length = 2.3 κ_s, and C_v ranges from 1 to 4 depending on the illumination of the object. The range of results shown in Figure 8-6 were developed from various experiments using people to sight objects through smokes from different fires. The results may not be so accurate but are very useful. Returning to our example for the PS, the smoke visibility is

$$L_v \approx (4,1)/49.9 \text{ m}^{-1} = 0.020 \text{ to } 0.080 \text{ m}$$

or about 2 to 9 cm: This is poor visibility. Note for the wood fire at $\Phi = 0.87$,

$$\kappa_s = \frac{(11 \text{ g/m}^2\text{s})(15.2 \text{ m}^2)(0.037 \text{ m}^2/\text{g})}{\left(\frac{(500+(11)(15.2))}{10^3}\right)\text{m}^3/\text{s}} = 9.27 \text{ m}^{-1}$$

and

$$L_v = \frac{(4, 1)}{9.27 \text{ m}^{-1}} = 0.108 \text{ to } 0.43 \text{ m}$$

This wood smoke is somewhat clearer, but visibility is still poor.

Heat Effects

Temperature can also cause harm by convective and radiative heat transfer. Two effects are possible: heat stress is relatively long term; a burn injury is more immediate. If skin reaches 45°C, we feel pain. At higher temperatures, damage can occur to increasing tissue depths. Figure 3-10 shows the time response for blister burns due to radiant heating of bare skin. We need at least 4 kW/m^2 to cause such burns in a relatively short time (in contrast to long-term exposure to the sun at ≤ 1 kW/m^2, which can result in a relatively mild burn). Purely convective heat fluxes associated with this heat flux threshold for a blister correspond to a given smoke temperature. This can be estimated from Equation (3-2), our formula for convective heat transfer. Selecting a heat transfer coefficient of h = 10 W/m^2-°C, approximating a stationary person, gives

$$\dot{q}'' = h(T_{\text{smoke}} - T_{\text{skin}})$$

$$T_{\text{smoke}} = \frac{\dot{q}''}{h} + T_{\text{skin}}$$

Solving,

$$= \left(\frac{4 \text{ kW/m}^2}{10 \text{W/m}^2 \text{-°C} \times 10^{-3} \text{ kW/W}} \right) + 45\text{°C}$$

$$= 445\text{°C}$$

Similarly, the pain threshold temperature for a stationary person is roughly 200°C. Actually, people have been exposed in tests to such high temperatures. The ability to perspire increases their tolerance time; after long heating the body core temperature can rise to cause heat stress in spite of perspiration. Heat stress occurs when the body's core temperature reaches 41°C; normal core temperature is 37°C. At 41°C, consciousness can be affected and further increases can lead to death by **hyperthermia**. Heat stress occurs when the body can no longer regulate its core temperature. This is roughly described by a balance of energies:

hyperthermia
heat stress

■ NOTE
Heat stress occurs when the body can no longer regulate its core temperature.

[Rate of energy stored
in body] = [Metabolic energy release rate]
+ [Radiative and convective heat rates]
– [Evaporation energy loss rate due to perspiration]
– [Respiration energy loss rate]

If the left-hand side is not zero, we have an imbalance, and the core temperature can go up (or down), indicating the start of heat stress. Table 8-5 lists some tolerance limits of people while at rest exposed to high temperatures. Notice that the relative humidity, RH, has an effect on the evaporation of perspiration, and therefore affects the tolerance time.

Table 8-5 *Tolerance times under heat stress conditions.*

Exposure Temperature °C	RH %	Tolerance Time
49	10	~10 days
49	50	~2 hours
49	100	~10 minutes
100	0–100	~10 minutes

Source: From Purser, Ref. 2.

Summary

Concentrations of combustion products can be computed from their yields. Yields are the fraction of combustion product produced per mass of fuel supplied. These yields depend on the fuel and on the fire condition: flaming conditions of over- or underventilation, smoldering, or heating. Concentration is the fraction of the product in the smoke. The effect of these concentrations over time can cause harm and may present a hazard. Usually dose, the product of concentration and time, is a direct measure of harm. Various criteria are available to quantify the level of harm. A summary of the primary effects to humans from combustion products is listed below:

Product	Effect
Temperature	Heat stress, burns
CO_2	Increase in respiration
Soot, tars	Visibility
O_2, CO, HCN	Loss of oxygen supply to blood
HCl, HF, HBr	Sensory irritant

Review Questions

1. A smoldering fire occurs in a room 40 m^3 in volume. The smoke remains in the room and maintains an average density of 0.9 kg/m^3. It is known for this case that the smoldering is steady at 1 g/s. It produces 0.2 g of carbon monoxide per g of fuel smoldered.
   a. Compute the mass of smoke gas in the room.
   b. Compute the mass of CO produced after one hour.
   c. Compute the mass (fraction) concentration of CO in the room after one hour.
   d. For a person resting, compute the COHb (%) using Equation (8-7) and Table (8-3). Note: The mass fraction (Y) of CO is nearly identical to its volume fraction (X).
2. In a 40-m^3 room the mass optical density (D_m) of the smoke is 0.33 m^2/g (from Table 8-1). For the 1 g/s smoldering fire, estimate the smoke visibility in the room at 1 hour, assuming smoke is well mixed throughout the room. Use Equation (8-9b) and Figure 8-6 for your estimation.
3. Will acetylene produce more soot than propane? Explain the basis of your answer.

True or False

1. Yield is synomomous for concentration.
2. An exposure to 50 ppm ($50/10^6$) of carbon monoxide can kill you.
3. Visibility through smoke does not depend on lighting. That is why exit signs are not lit.
4. Acid gases, such as hyrogen chloride, can cause corrosion of computer equipment after a fire.
5. Heat stress is usually not a factor to people trapped in a fire, but is a serious problem for firefighters.

Activities

1. Examine fire incident statistics and draw from your experience to assess the time span involved in lethal smoldering fires.
2. Why are rats or mice used in smoke toxicity testing to assess the hazard to humans? Do their results directly apply to humans? You will need to research this answer.

References

1. A. Tewarson, "Generation of Heat and Chemical Compounds in Fires," chap. 3-4 in *SFPE Handbook of Fire Protection Engineering,* 2d ed., edited by P. J. DiNenno (Quincy, MA: National Fire Protection Association, June 1995).
2. D. A. Purser, "Toxicity Assessment of Combustion Products," in chap. 2, sec. 8 in *SFPE Handbook of Fire Protection Engineering,* 2d ed., edited by P. J. DiNenno (Quincy, MA: National Fire Protection Association, June 1995).
3. J. G. Quintiere, "Smoke Measurements: An Assessment of Correlations between Laboratory and Full-scale Experiments," *Fire and Materials* 6, no. 3 and 4 (1982): 145–160.

Compartment Fires

Learning Objectives

Upon completion of this chapter, you should be able to:

- Explain the processes in the development of fire in a compartment.
- Explain flashover, fully developed, and ventilation-limited fire in a compartment.
- Explain fire-induced flows, neutral plane, and ventilation factor.
- Compute vent flow rates and compartment smoke temperatures.

INTRODUCTION

developing fire
the early stage of growth (in a compartment fire) before flashover and full involvement

flashover
a rapid change in a developing room fire to full room involvement

backdraft
the sudden eruption of fire in a compartment due to the introduction of fresh air

Fire in a compartment involves the containment of smoke and the potential for the fire to spread beyond the compartment. Small fires can transport smoke great distances. Large fires can be affected by the compartment: enhanced due to thermal feedback (increased heat transfer) and diminished due to oxygen vitiation (decreased air flow to the fire). The motion of smoke and air in these fires is mainly due to buoyant effects as a result of the increased temperatures. Typically, the forced conditioned air supply in a building is secondary to this buoyant flow. We examine these effects and introduce some simplified, but very useful, formulas for assessing smoke temperatures and flow rates.

STAGES OF FIRE DEVELOPMENT

Fire in a compartment initially is unaffected by the compartment conditions. But as the oxygen is reduced in the compartment and as the compartment temperature rises, the compartment conditions play a role in the behavior of the fire. The stages of a fire in a compartment can be described as follows. Smoldering fires are excluded and discussed later.

NOTE
Flashover usually causes a fire to reach its fully developed state in which all of the available fuel becomes involved.

1. **Developing fire:** Following ignition, the fire grows on an item and may also involve other items. The oxygen concentration and compartment temperatures are not much different than normal air. The fire behaves as if it is burning in the open air for most of this period. See Figure 9-1a.
2. **Flashover:** Flashover marks a dramatic increase in fire conditions due to the confinement of a room. It can be caused by several situations:
   a. The rapid ignition and flame spread of materials due to heat flux increasing.
   b. The accumulation of fuel rich gases and their sudden exposure to air, commonly called **backdraft** by the fire service.
   c. The increase in the burning rate and the sudden extension of flames through the room.

ventilation-limited or ventilation-controlled
state of a compartment fire where the air supply is limited; smoke gases will have nearly zero oxygen left; underventilated

fuel-limited
state of a compartment fire where the air supply is sufficient to maintain combustion

Situations a and c refer to the depiction in Figure 9-1b.

Flashover usually causes the fire to reach its fully developed state in which all of the fuel within the room becomes involved. However, all of the fuel gases may not be able to burn within the room because the air supply is limited. Such an air-limited fire is termed **ventilation-limited** or **ventilation-controlled**, as opposed to a **fuel-limited** fire, which is indicative of the developing fire. Previously for this air-limited state we used the term *underventilated* when discussing yield data. For compartment fires, these new terms have become common usage. When the ventilation-limited condition occurs, the production of CO, smoke, and energy are at their highest values. Also, the oxygen concentration in the smoke layer becomes nearly zero. Flashover marks a transition in which the fire devel-

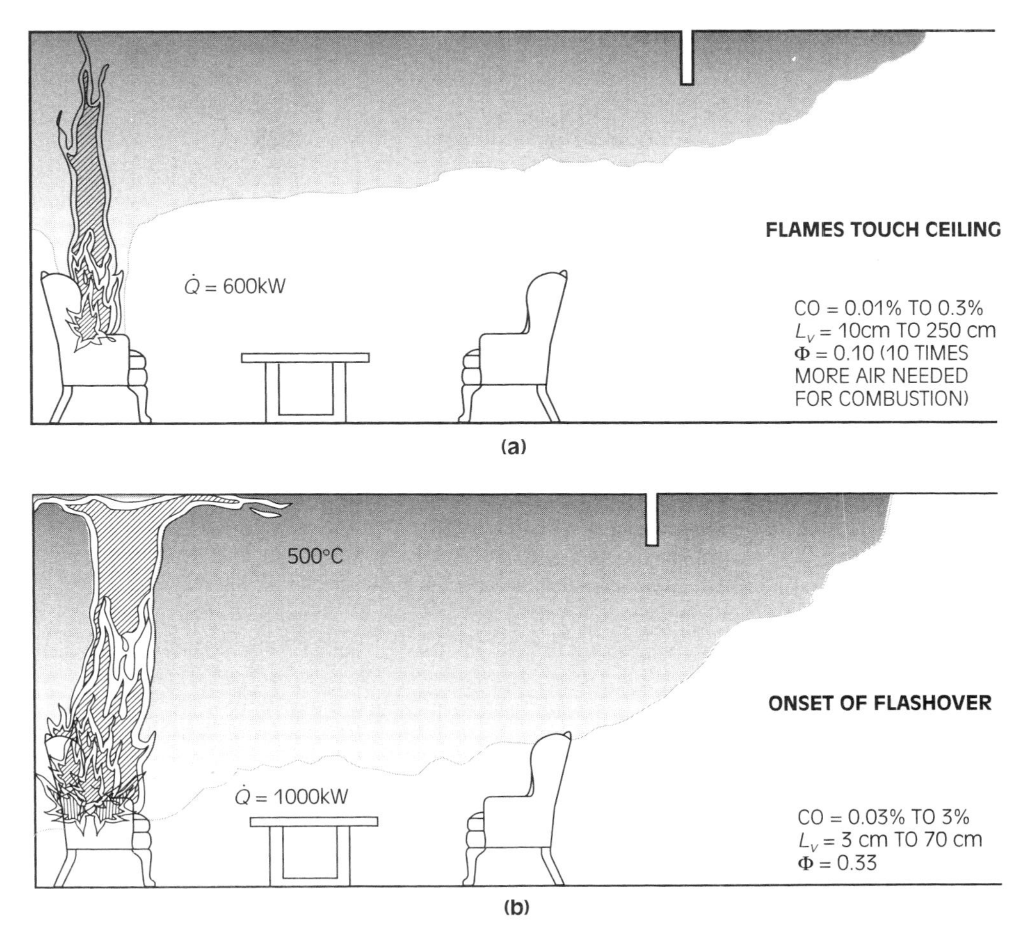

Figure 9-1 *Typical room conditions for furniture items burning in a 10 × 10 × 8 ft high room with a 3.2 × 6.4 ft high open doorway. CO = concentration of carbon monoxide, L_v = smoke visibility, Φ = equivalence ratio, Φ > 1 indicates unburned fuel.*

opment was previously dominated by the fuel materials and subsequently is primarily controlled by the ventilation conditions indicative of the room and building geometry.

fully developed state of a compartment fire during which the flames fill the room involving all the combustibles

3. Fully Developed: This stage is marked by flames fully encompassing the room, with the likelihood of flames emerging from windows and doors. All of the fuel is involved to its maximum potential. This period involves possible structural damage, and heat flux conditions in the room can reach as high as 150 kW/m^2. Depending on the available air supply, this stage may or may not be ventilation limited. Most compartments and buildings become ventilation limited at this stage. A fire in a glass-tower office building, with full fire on a floor and all the windows broken, could be fully developed but not ventilation limited. For a ventilation-limited fire, the room's energy release rate, $\dot{Q}$, is established by the available air supply rate. See Figure 9-1c.

An example of the variables in the smoke contained in a developing room fire is shown in Figure 9-2. This fire was initiated as a smoldering fire that changed to flaming at about 18.5 min. Flashover is perceived at 25.5 min. The temperature is 500°F (260°C) at that instant, but reaches about 1,000°F (538°C) in less than a minute. This also demonstrates that the perception of flashover seen by an observer is not necessarily that which correlates precisely with 500° to 600°C. Therefore this temperature criterion is somewhat displaced by about 30 s. Such indicators and criteria for the onset of flashover are subject to variation. But there is no doubt that a sudden transition has occurred. The O_2 drops to zero, the CO peaks at 7%, and the temperature ultimately reaches 1,600°F (870°C). Typical, fully developed compartment fire temperatures reach 800° to 1,000°C. Note that the

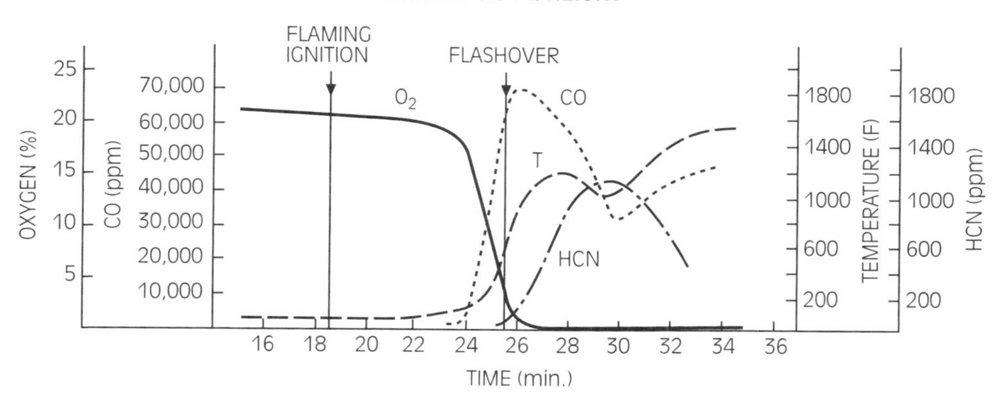

Figure 9-2 *Furnished room fire initiated by a smoldering upholstered chair. After Hartzell, Ref. 1.*

HCN concentration from the polyurethane foam chair does not increase until well after flashover. Also, the fact that the oxygen drops to zero and remains there clearly shows the fire has used all of the oxygen in the air that entered the room and is ventilation limited.

FIRE-INDUCED FLOWS

Flows occur as a result of fire, or in general due to any heat source through buoyancy. Buoyancy causes a pressure difference that results in a velocity. As an extension of the concepts in Chapter 7, this pressure difference Δp (Δ means a difference in p, i.e., $p_2 - p_1$), is

$$\Delta p = (\rho_a - \rho) g H \quad \textbf{(9-1)}$$

where ρ_a is the air (cold) density,
ρ is the smoke of fire (hot) density,
g is the gravitational acceleration, 9.81 m/s^2, and
H is a vertical height (between 1 and 2).

In Chapter 7, we called this relative potential energy per unit volume:

$$\frac{\text{Energy}}{\text{Volume}} \text{ in } \frac{\text{J}}{\text{m}^3} \sim \text{Pressure in } \frac{\text{Newton (N)}}{\text{m}^2}$$

One Newton (N) is the same as 1 kg-m^2/s, and 1 Joule (J) is 1 kg-m^3/s or 1 N-m; therefore

$$1\frac{\text{N}}{\text{m}^2} = 1\frac{\text{kg - m}^2/\text{s}}{\text{m}^2} = 1\frac{\text{kg - m}^3/\text{s}}{\text{m}^3} = 1\frac{\text{J}}{\text{m}^3}$$

Newton's second law of motion
relates force on a body to its mass and resulting acceleration

These are equivalent units of measure although their names are different. They are related through **Newton's second law of motion**: Force = Mass × Acceleration. Accordingly, the resultant velocity is

$$V = \sqrt{\frac{2(\rho_a - \rho) g H}{\rho}} \quad \textbf{(9-2)}$$

which follows from Equation (7-5). From Equations 9-1 and 9-2 we can obtain approximate formulas to explain the flows associated with fire. They should not be used for quantitative calculations, only for explanations. Therefore, we shall use ~ meaning "depends on," rather than = which means "equal to."

Let us roughly explain the relationship of fire-induced flows compared to fan-generated flows. Such fan flows would occur in air handling systems. Consider two cases as shown in Figure 9-3. The first is pure fan flow over height H in a duct of diameter D. The second is purely buoyant flow where the duct gases are at tem-

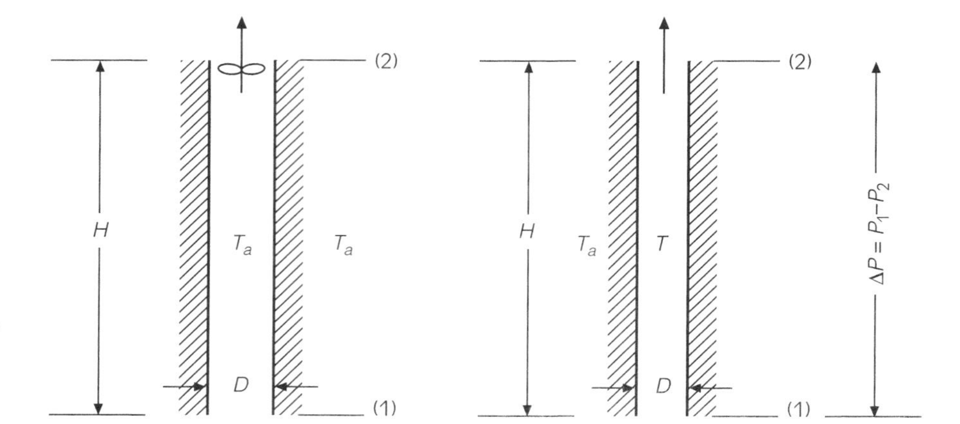

Figure 9-3 *Ideal fan and buoyancy flows in a duct.*

perature, T, compared to the air at T_a. The fan would create a pressure difference to just balance the friction effects as given by

$$\Delta p_{\text{fan}} = \text{f}\frac{H}{D}\frac{V^2}{2} \tag{9-3}$$

where f is a friction factor.

The buoyant duct would create a pressure difference given by

$$\Delta p_b = (\rho_a - \rho)gH$$

or

$$\Delta p_b = \rho_a(1 - T_a/T)gH \tag{9-4}$$

If we assume the fan needs to supply flow at 0.1 m/s in a $D = 0.3$ m duct, then

$$\Delta p_{\text{fan}} \sim 0.167\text{f}\, H$$

and typically $\text{f} \approx 0.001$ to 0.07, so that

$$\Delta p_{\text{fan}} \sim 0.017H \text{ to } 0.11H \frac{\text{N}}{\text{m}^2}$$

In comparison,

$$\Delta p_b \sim (1.2\ \text{kg/m}^3)(1 - 0.5)(9.8\ \text{m/s}^2)\ H$$

where we have assumed $T_a \sim 300$ K and $T \sim 600$ K. Then, $\Delta p_b \sim 5H$ N/m^2. The fan-generated pressure increase would approximately equal the buoyancy pressure at roughly a supply velocity of 6 m/s in this example.

It is clear that, under fire conditions, buoyant flows are likely to dominate fan-generated duct flows. Also, if $H = 1$ m, $\Delta p_b \sim 5$ N/m^2, compared to normal atmospheric pressure of 10^5 N/m^2. The pressures due to fire and also due to air handling systems do not cause any significant change in the overall building pressure. Very, very small pressure differences drive these flows, and the overall building pressure does not significantly depart from 10^5 N/m^2.

COMPARTMENT FLOW DYNAMICS

Let us consider how smoke flows in a single room and in a building.

Layers and Vent Flows

A fire in a room with an open door or window vent would first cause smoke to gather at the ceiling in a distinct layer. Then the smoke layer would drop below the vent soffit and begin to flow out of the vent. The smoke layer would continue to drop as the temperature, or fire $\dot{Q}$, increases. Cooling or a decrease in $\dot{Q}$ causes the layer to rise. A fully developed room fire can have a layer several inches above the floor. This scenario is depicted in Figure 9-4. Flow enters the vent below the neutral plane, and smoke leaves above this height. The smoke layer is

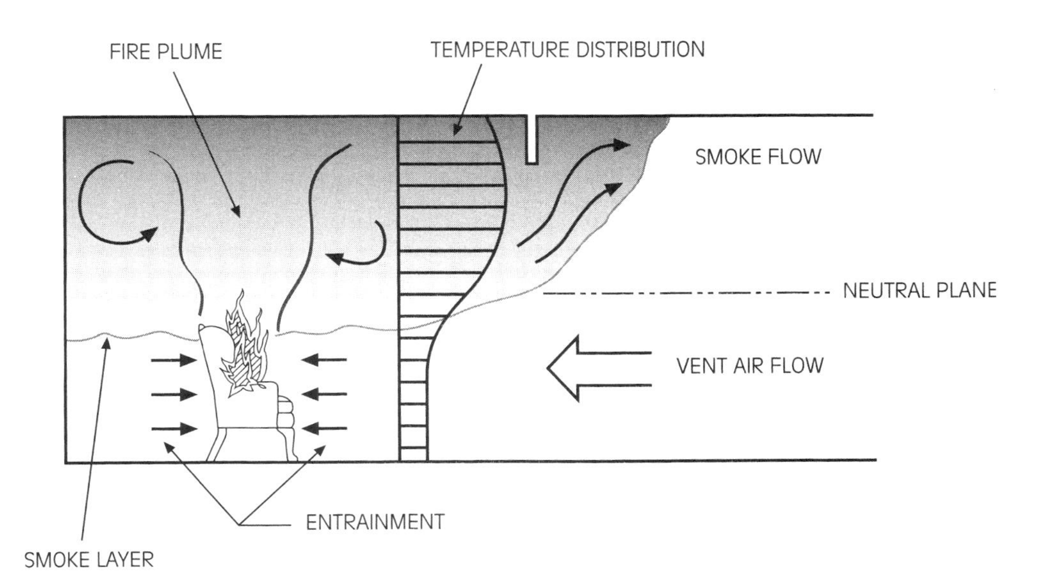

Figure 9-4 *Fire-induced flows in a room.*

lower than the neutral plane in the room. Over this layer height entrainment supplies air to the fire. Very little mixing occurs across the smoke layer interface. Actual temperature data for the room gases and wall surfaces are shown in Figure 9-5. In general, the smoke layer gets stirred, resulting in a nearly uniform temperature outside of the fire plume. Some mixing and heating gives the lower layer a temperature above the entering air. Note how the wall temperatures behave in Figure 9-5. These flow dynamics give rise to an approximation used to predict compartment fires—a two-layer well-mixed model. In any case, this two layer concept is a good way to view smoke behavior in fire. It is stably stratified, that is, its higher temperature keeps it above the relatively cooler air. Very little mixing occurs between these hot and cold gas regions. But mixing can occur at vents, at cooler walls, in shafts, etc. The cold gas region can become heated and contaminated with combustion products.

Smoke Filling

If a compartment is closed or in the early stage of the fire before smoke leaves through vents, the dynamics are those of filling. This process is the inverse of fill-

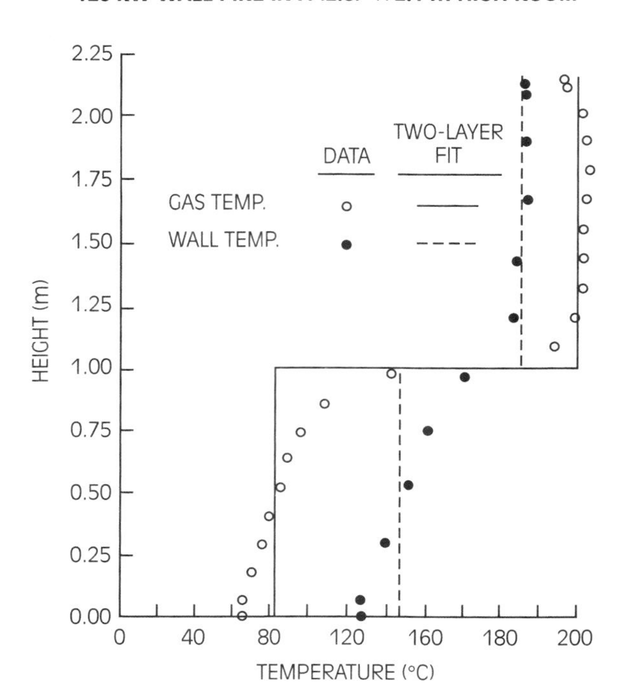

Figure 9-5 *Room temperatures with a wall vent at steady state. After Quintiere et al., Ref. 2.*

■ **NOTE**
If a fire dies due to lack of oxygen, cooling occurs and the pressure becomes negative, drawing in air that can revitalize a fire, starting the breathing again.

ing a tub with water. The smoke interface moves down to fill the container; and even if the container has a leak, the container will fill if the fire is sufficient. All buildings leak; they are not made to be perfectly sealed. Some exceptions are nuclear reactor containment buildings, spacecraft, and some industrial containers. In these, a fire causes the pressure to rise, and their containers may then structurally fail. For normal buildings, the pressure changes slightly to cause the appropriate flow. The filling of a compartment with floor and ceiling leaks is shown in Figure 9-6. During filling and leaking, the pressure remains slightly positive causing flow out of the room. If the fire dies due to lack of oxygen, cooling occurs and the pressure becomes negative, drawing air in. This air can revitalize the fire, and the breathing starts again. Firefighters know this sign and avoid venting such "breathing" enclosures. This situation has a potential for back draft.

A room with a vent fills in the early growth stage. The flow and pressure dynamics are illustrated in Figure 9-7. Once the smoke layer interface, H_L, falls below the vent soffit, the flow in the vent begins to become bidirectional. Thereafter, the flow rate into the room is nearly equal to the flow rate out. For a constant smoke temperature, Figure 9-7c prevails. The **neutral plane** height may change from roughly $H_o/2$ to $H_o/3$ as the compartment becomes fully involved in flames. At that event the smoke layer can drop to as little as inches off the floor.

neutral plane
the height above which smoke will or can flow out of a compartment

Smoke Movement in a Building

The dynamics of smoke movement through a building have not been sufficiently studied, however, we can qualitatively describe the smoke's behavior.

Figure 9-8 shows steady flow into a corridor that has an exit vent. A stratified smoke layer whose interface height depends on the size of the exit opening is formed in the corridor. The smaller the exit opening, the lower is the interface (see Figure 1-5, page 15). Within the corridor hot and cold layers, the flow pattern is complex as shown in Figure 9-8 (or Figure 1-6, on page 16).

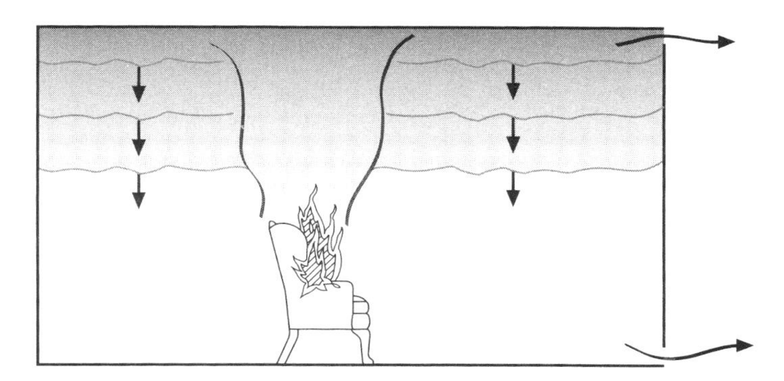

Figure 9-6 *Smoke filling a leaky space.*

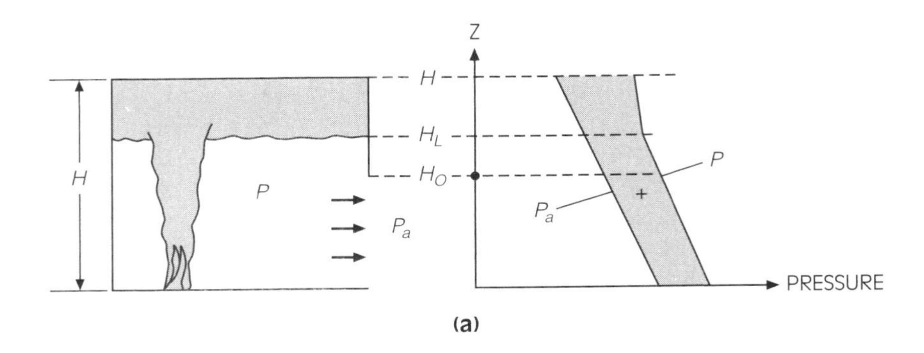

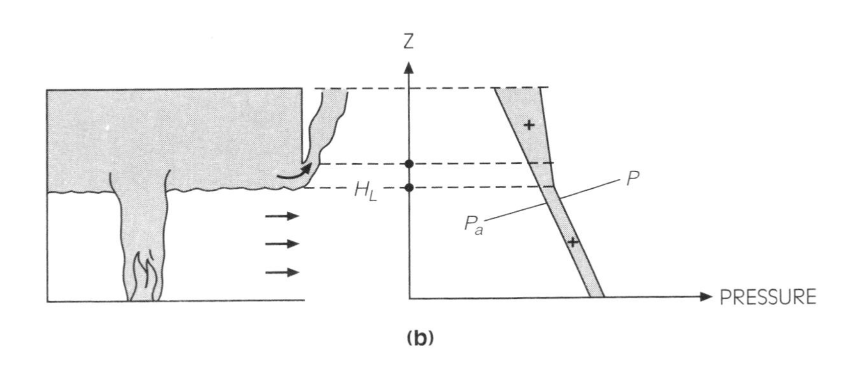

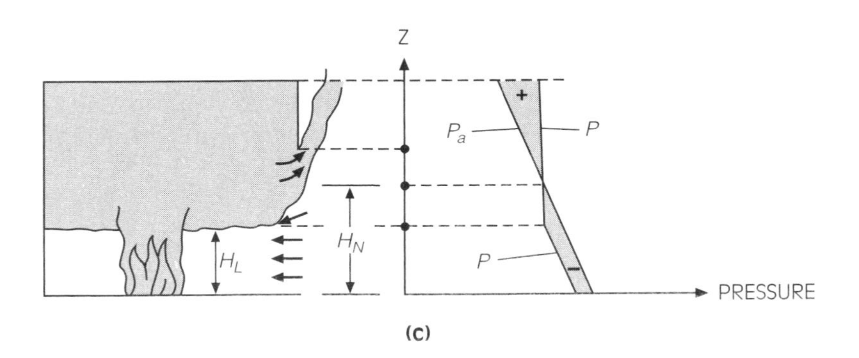

Figure 9-7 *Flow and pressure dynamics in a room with a vent. (a) Layer descending; cold flow leaves the vent. (b) Hot flow begins to leave the vent. (c) Layer interface below the vent and cold flow enters from the surroundings.*

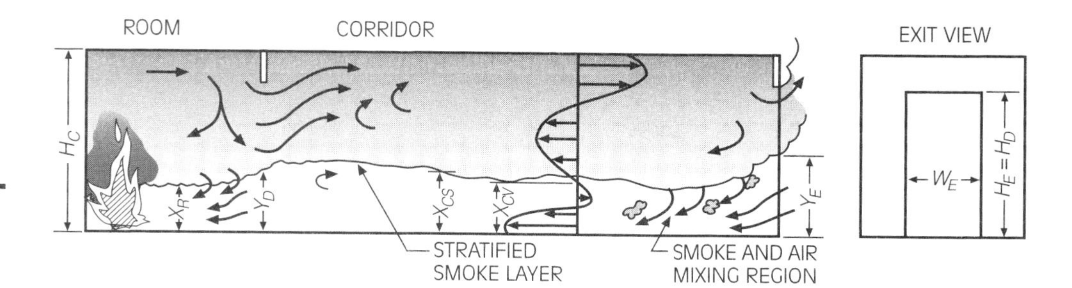

Figure 9-8 *Fire-induced flows in a room and corridor.*

If the corridor has no exit vent, it will fill, much like a single compartment, as smoke emerges from a room. If the flow is gradual, we have the process shown in Figure 9-9a and in the photographs of Figure 9-10a–d. In 9-10d, the smoke begins to dribble down the cooler walls as the layer interface approaches the floor. Another possible flow is due to a sudden release or increase of smoke into the corridor. The initial momentum, kinetic energy, of the smoke could cause a wave front progation that could turn back on itself. Such a flow would fill the corridor by a lateral motion, compared to the more tranquil filling (see Figure 9-9b). The velocity of this front is somewhat less than that given by Equation (9-2), with H given as the corridor height.

Flow up a tall shaft, such as an elevator conduit, would initially be plume-like until the walls intercept the plume. This flow is shown in Figure 9-11. After interception, a complex mixing process occurs that carries the smoke upward. Once the shaft is filled with hot smoke it acts like any other chimney. The subsequent flow in the shaft depends on vents and leakage in the shaft.

Figure 9-12 shows the flow processes in a building having been filled with smoke. The hot shaft carries the smoke to the upper floors, and smoke leaks out above the neutral plane of the building. For the simple case of equal top floor and fire floor vents, the height of the building neutral plane relative to the fire floor is

$$\frac{H_N}{H_B} = \frac{1}{1 + T_a / T} \tag{9-5}$$

where T_a is the surrounding air temperature in K and T is the average smoke temperature in K.

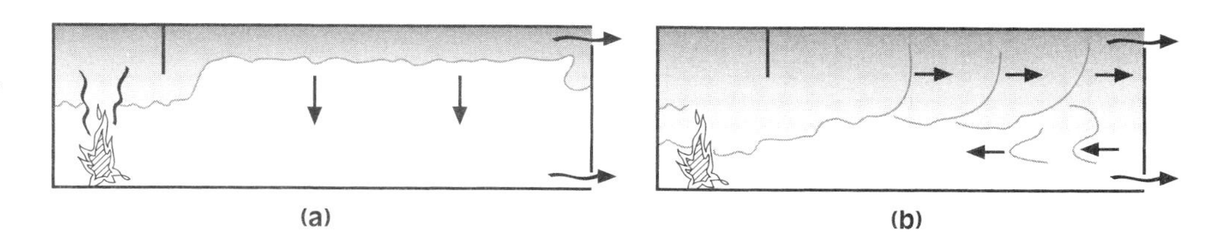

Figure 9-9 *Smoke filling a closed corridor (a) slowly, (b) impulsively.*

Figure 9-10 *Smoke filling a closed corridor from a room fire on the right. (a) Incipient smoke impinging on the left wall of the corridor.*

Figure 9-10 (continued)
(b) Smoke spreading along the corridor ceiling.

Figure 9-10 (continued)
(c) Smoke layer descending mid-height.

Figure 9-10 (continued) *(d) Smoke layer descending near floor.*

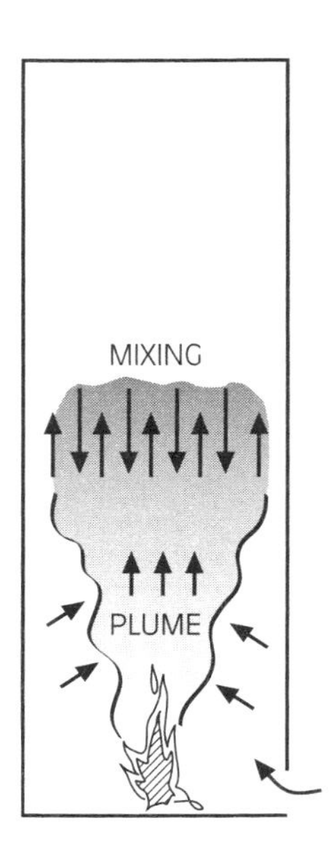

Figure 9-11 *Smoke filling a shaft.*

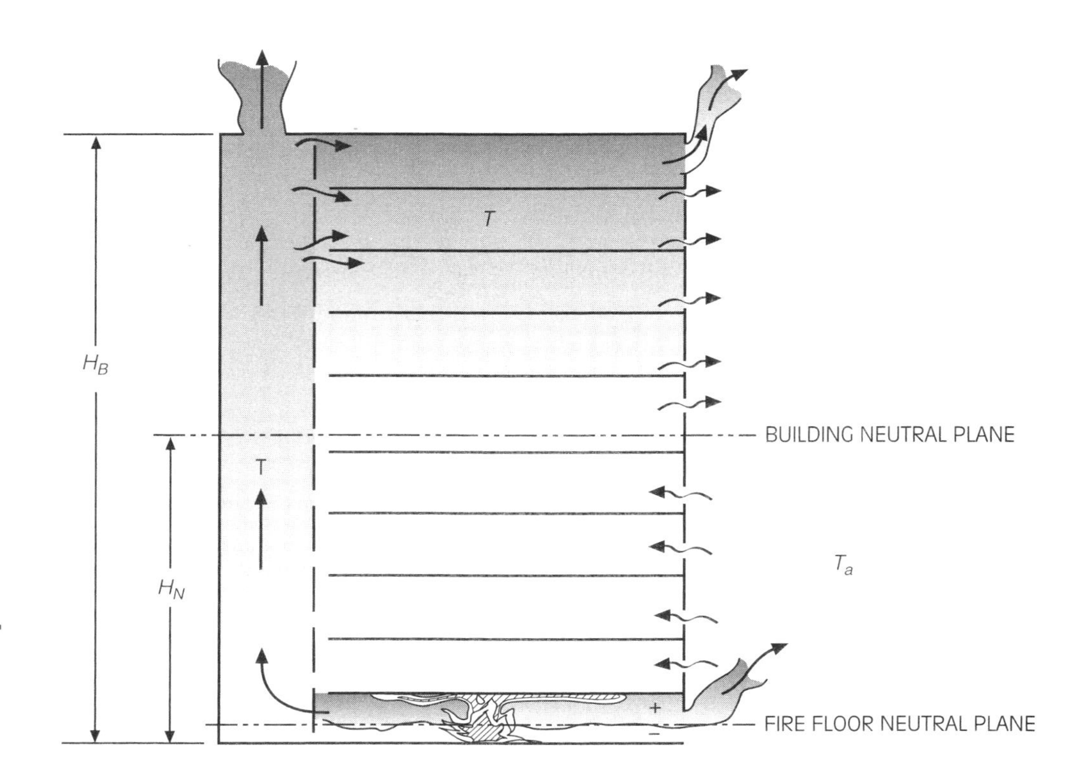

Figure 9-12 *Simplified dynamics of smoke filling a building.*

Since the smoke would cool considerably in a large building, $T \approx T_a$. It is not unreasonable to find $H_N/H_B = 0.5$. Note that the fire floor would have its own neutral plane, and we have ignored this in arriving at Equation (9-5). Also wind, natural building flows due to buoyancy, and flows due to a building's air-heating or cooling system affect the building smoke dynamics. These natural flows due to the normal buoyancy of the building air are referred to as **stack effect** flows. A cool building relative to the environment displays flows opposite to Figure 9-12. An early fire in an air-conditioned building in the summer would experience downward motion of smoke and leakage of smoke below the fire floor.

stack effect
buoyancy of the air in a tall building due to normal air-conditioning and heating systems.

SINGLE ROOM FIRE ANALYSIS

Let us consider how to make some quantitative fire estimates for a single room. We consider a room with a vent and show how to estimate the smoke flow rate through the vent and its temperature.

Vent Flows

We have seen in Figure 9-7c that the pressure difference caused by the difference in smoke and air temperatures drives the flow across the vent. The rate of smoke flow out is nearly equal to the rate of air flow in. The exact rates depend on the temperature difference and the size of the vent opening. For a vent of height H_o and cross-sectional area of A_o the maximum air flow rate, $\dot{m}_{a,\max}$ is given by

$$\dot{m}_{\mathrm{a,max}} = 0.5 A_o \sqrt{H_o}\ (\mathrm{kg/s}) \tag{9-6}$$

where A_o is in square meters and H_o is in meters.

Large vents and temperature well below 800°C have lower flow rates, as illustrated by actual data shown in Figure 9-13 where the theoretical maximum, Equation (9-6), places an upper bound on the data. For large fires and small vents, Equation (9-6) is a good approximation in estimating the air supply rate. Incidentally, $A_o\sqrt{H_o}$ is called the **ventilation factor** because it is the principal variable controlling flow. A_o accounts for the flow area and the $\sqrt{H_o}$ accounts for the velocity as described by Equation (9.2).

ventilation factor
the parameter controlling smoke flow rate through a door or window

SMOKE TEMPERATURE

We would expect the average temperature of the smoke layer, not counting the plume region, to be affected by the size of the fire, $\dot{Q}$, and the ventilations factor, $A_o\sqrt{H_o}$. The larger the fire $\dot{Q}$ and the smaller the vent, the higher we expect the smoke temperature. It is not surprising an approximate formula for the smoke temperature rise $\Delta T = T - T_a$ is

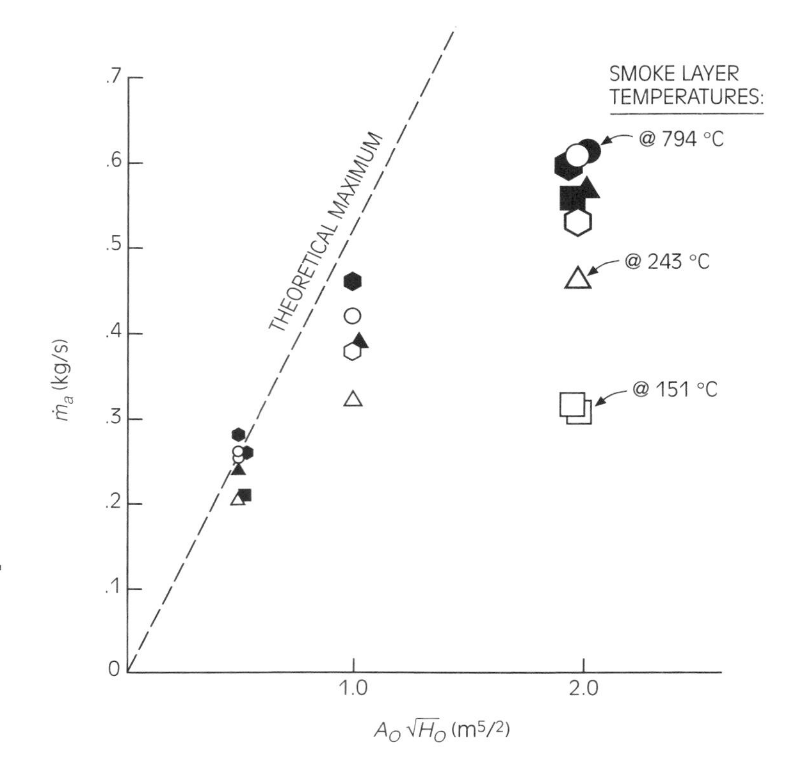

Figure 9-13 *Typical air flow rates through vents in room fires. After Quintiere at al., Ref. 3.*

$$\Delta T\,(^\circ C) = 6.85\left[\frac{\dot{Q}^2}{(hA)(A_o\sqrt{H_o})}\right]^{1/3} \qquad \textbf{(9-7)}$$

where A_o is in square meters,
H_o is in meters,
A is the interior surface area, in square meters, and
h is a heat loss coefficient taken as the larger of

$$h = \text{Maximum of}\left(\sqrt{\frac{k\rho c}{t}} \text{ or } \frac{k}{l}\right) \text{ in } \frac{\text{kW}}{\text{m}^2 \text{ - }^\circ\text{C}}$$

thermal inertia
a thermal property responsible for the rate of temperature rise

where $k\rho c$ is the interior construction **thermal inertia** $\left(\frac{\text{kW}}{\text{m}^2 \text{ - }^\circ\text{C}}\right)^2$ – s,
k is thermal conductivity $\frac{(\text{kW})}{(\text{m - }^\circ\text{C})}$,

l is the interior construction thickness, in meters, and
t is time in seconds. Table 9-1 gives typical values of k and $k\rho c$ for common construction materials.

Both Equations (9-6) and (9-7) can be used for multiple vents by summing the $A_o\sqrt{H_o}$ values and can be used for different construction materials by summing the hA values for the various wall, ceiling, and floor elements.

For example, consider a room 4 × 4 × 3 m high with a simple vent 1 × 2 m high. The construction is essentially gypsum plaster with properties given in Table 9-1. The fire is constant at 500 kW. Assume the gypsum is very thick. Compute the temperature rise of the smoke at 100 s.

$$A_o = 2 \text{ m}^2$$

$$A_o\sqrt{H_o} = 2\sqrt{2} \text{ m}^{5/2}$$

$$A = 2[4\times4+4\times4+3\times4]-2 = 86 \text{ m}^2$$

$$h = \sqrt{\frac{k\rho c}{t}} = \sqrt{\frac{0.60}{100}} = 0.0775\frac{\text{kW}}{\text{m}^2-°\text{C}}$$

Therefore, from Equation (9-7)

$$\Delta T = 6.85\left[\frac{(500)^2}{(0.0775)(86)\left(2\sqrt{2}\right)}\right]^{1/3}$$

$$= 6.85(26.6)$$

$$\Delta T = 162°\text{C}$$

Table 9-1 *Typical construction properties.*

	$k\rho c$ ($kW^2 \cdot s/m^4 \cdot °C^2$)	**k (kW/m·C)**
Insulation board	**0.09**	**4.1×10^{-5}**
Wood	0.30	1.5×10^{-4}
Gypsum board	0.60	5×10^{-4}
Concrete	2.0	1×10^{-3}
Steel	150	5×10^{-2}

Flashover

Let us now ask what conditions would lead to a temperature rise of 500°C. Black-body radiant heat flux at this temperature (500° + 25°C air temperature) is 23 kW/m^2 (from σT^4). At this heating level we can expect other items to ignite. Such ignitions or fire growth could be perceived as the onset of flashover. Thus, if we use a ΔT = 500°C to indicate flashover potential, substitution in Equation (9-7) gives the $\dot{Q}$ required for flashover to start:

$$\dot{Q}_{FO} = \left[\frac{(500)}{(6.85)}\right]^{3/2}\left[(hA)\left(A_o\sqrt{H_o}\right)\right]^{1/2}$$

or

$$\dot{Q}_{FO} = 624\left[(hA)\left(A_o\sqrt{H_o}\right)\right]^{1/2} \quad \textbf{(9-8)}$$

For the previous example, presuming flashover occurs near 100 s,

$$\dot{Q}_{FO} = 624\left[(0.0775)(86)\left(2\sqrt{2}\right)\right]^{1/2}$$
$$= 2{,}709 \text{ kW or } 2.71 \text{ MW}$$

Ventilation-Limited Fires

Because we know how to compute the maximum air flow rate, which is valid for small vents or large fires, we have for the above example,

$$\dot{m}_a = 0.5\left(2\sqrt{2}\right) = 1.41 \text{ kg/s}$$

Almost all common fuels produce approximately 3 kJ per gram of air (or oxygen in the air consumed). If all the oxygen is consumed we have a ventilation-limited fire and the maximum possible energy release rate within the room is

$$\dot{Q}_{max} = \dot{m}_{a,max} \text{ (kg/s)} \times 3{,}000 \text{ kJ/kg} \quad \textbf{(9-9)}$$

For our example, this is

$$\dot{Q}_{max} = 1.41 \times 3{,}000 = 4{,}243 \text{ kW or } 4.24 \text{ MW}$$

We see that this energy release is greater than the value to initiate flashover. This difference between flashover and ventilation-limited energy is true in general for most compartment fires.

Fully Developed Fire Size

Equation (9-9) gives the energy release rate within the room. The maximum fuel burning rate depends on what is there. Typically, fully developed fires are 800°C or higher, which corresponds to a radiant heat flux (σT^4, $T = 1073$ K) is 75 kW/m^2. From Chapter 6, we saw that the heat of gasification may range from 1 to 5 kJ/g. From Equation (6-1), assuming an average L of 2 kJ/g for the above example with the fuel distributed over the floor area A = 12 m^2:

$$\dot{m} = \frac{(75 \text{ kW/m}^2)}{2 \text{ kJ/g}}(12 \text{ m}^2) = 450 \text{ g/s}$$

If the average heat of combustion, ΔH_c, is taken as 20 kJ/g, then the corresponding fire size (Equation (6-2)) is

$$\dot{Q} = (450 \text{ g/s})(20 \text{ kJ/g}) = 9{,}000 \text{ kW or } 9.0 \text{ MW}$$

The remaining energy release rate, 9 – 4.24 = 4.76 MW, must potentially burn outside of the room, which shows the consequences of flashover. If instead of just considering floor fuel, we had said the enclosure was completely lined with fuel, e.g., combustible interior finish, then A = 86 m^2, and $\dot{m}$ = 3,225 g/s and $\dot{Q}$ = 64.5 MW. The hazard of combustible interior finish, even thin ones, is clearly demonstrated. Figure 9-14 shows the possible fire growth behavior of the room example. The

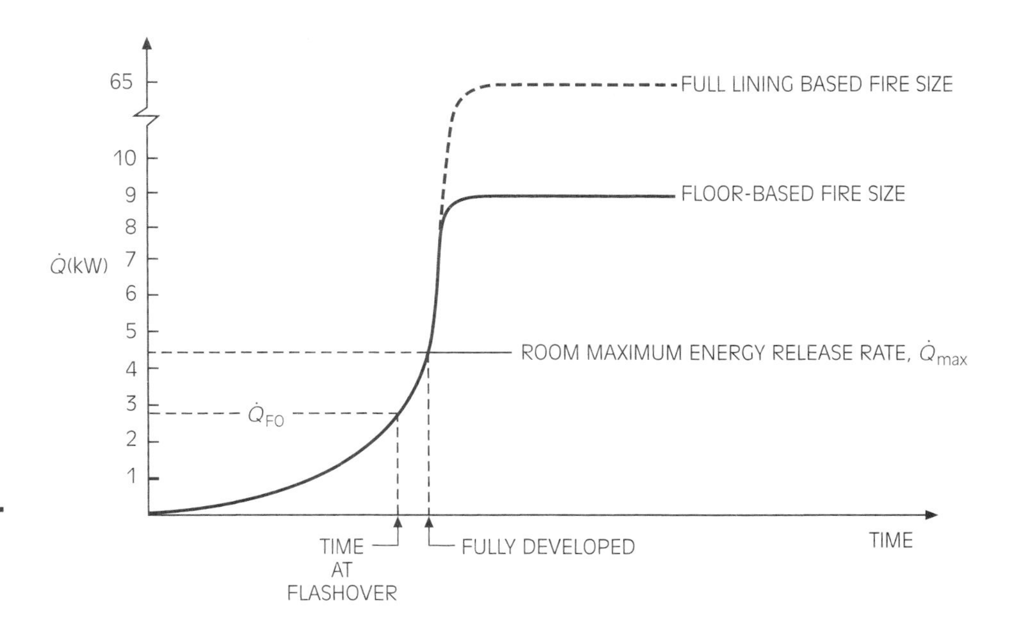

Figure 9-14 *Fire growth of example fire scenario.*

fully developed fire state depends on the nature and extent of the fuel available to burn.

SMOLDERING FIRE IN A CLOSED SPACE

Figure 9-15 shows the smoke dynamics of a smoldering upholstered chair fire in an 8.8 $m^2 \times 2.4$ m high closed room. The smoldering was initiated by a simulated cigarette lighter.

Figure 9-16 shows the corresponding CO concentration as the smoke layer descends. The smoke reaches the floor in 67 min. Typically, the yield of CO in such fires is 0.1 to 0.2 for cotton or flexible polyurethane upholstery. Oxygen remains high (typically > 16%) and temperature increases are small (<50°C).

Figure 9-17 shows theoretical results for this chair fire in 2.4 m high rooms of different sizes. The CO concentrations are computed, and the time to reach an incapacitation level (taken as CO (%) × time = 4.5%-min) is indicated as t^*. The smoldering source on the chair was placed at 0.4 m above the floor; and the CO is sensed at a height of 1.2 m, representative of a sitting position. The time for the smoke layer to first reach 1.2 m above the floor is indicated by t_o in Figure 9-16. For residential room size spaces, incapacitation occurs within 1 h, but for house size spaces it may take up to 2 h. These figures are consistent with actual smoldering fire incident data, but similar times are also found for the smoldering fire to make a transition to flaming. This transition is not currently predictable; it strongly depends on aeration and flow velocity at the smolder region. Both processes, flaming or continued smoldering, can be deadly to occupants in these times of 1 to 2 h. Smolder continuously yields low levels of CO, but a ventilation-limited fire can lead to CO concentrations of 2% to 7%. The latter can cause incapacitation in less than 2 min.

Figure 9-15 *Photo sequence of smoldering chair smoke filling a closed room.*

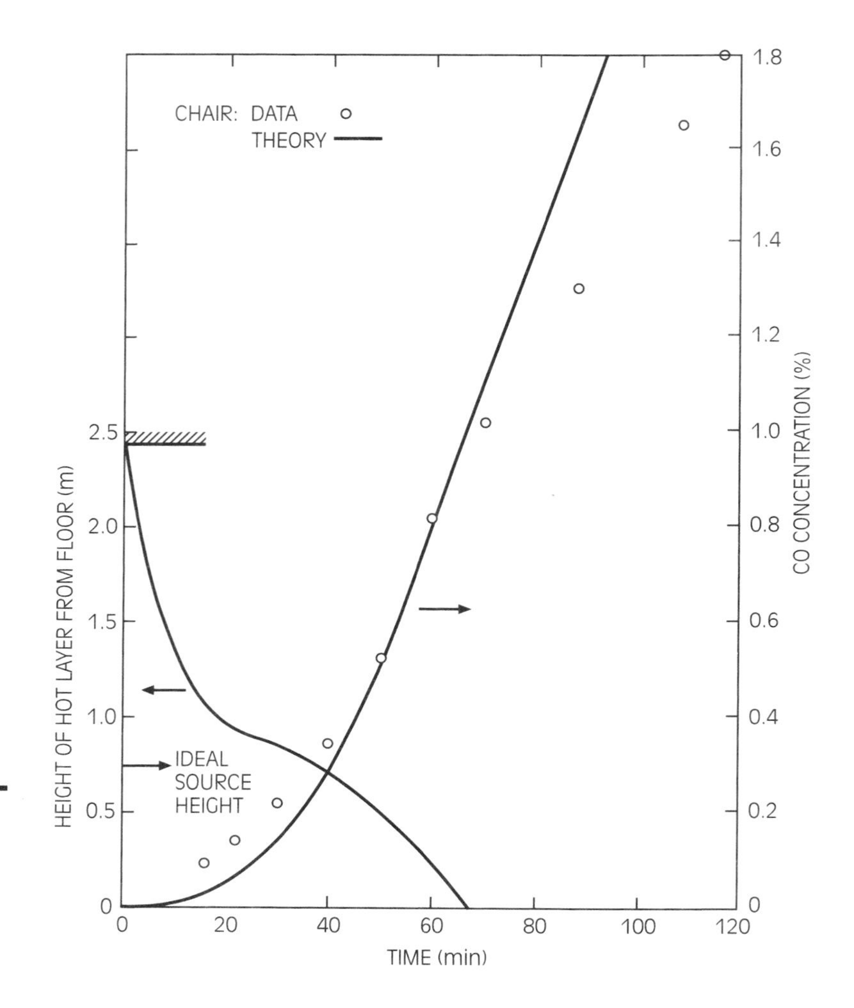

Figure 9-16
Dynamics of smoldering chair fire in closed room. After Quintiere et al., Ref. 4.

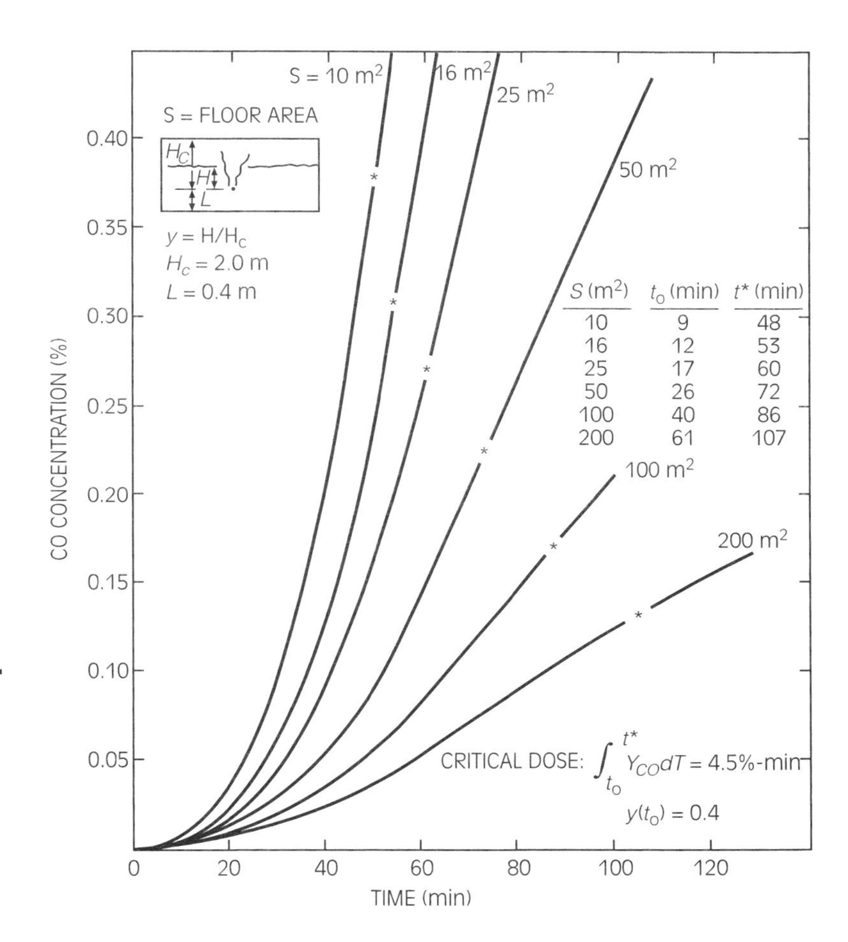

Figure 9-17 *The increase of carbon monoxide for a smoldering chair fire in rooms of different sizes. After Quintiere et al., Ref. 4.*

Summary

Fire grows in a room in stages: first, a nearly free-burning stage without any influence from the room; then flashover can occur; followed by the fully developed state, which is likely to be ventilation limited.

Smoke predominately propagates through a building due to buoyancy caused by the heat of the fire. Very small but distinct pressure differences drive the flows. The smoke generally stratifies in an upper hot layer.

Air flow rate through vents and smoke layer temperature in a ventilated room can be predicted by simple formulas, which can be used to describe many aspects of fire growth.

Smoldering fires, which produce significant yields of CO, can lead to incapacitation in roughly 1 to 2 hours in typical dwelling spaces.

Review Questions

1. For your classroom estimate the energy release rate needed to initiate flashover. Assume the door is open. What kind of fuel package might this represent?
2. What fire size (kW) will cause a 300°C smoke layer temperature in your room with the door half open? What is the maximum radiant heat flux associated with smoke of 300°C?
3. If your room reaches flashover and becomes fully developed reaching 800°C, what is the maximum air flow rate through your fully opened door? If all the windows break, what is the additional air flow rate? What fire size can be supported by these total air flow rates? Is there sufficient fuel in the room to do it?
4. The World Trade Center on a winter day has an indoor temperature of 23°C and an outdoor temperature of 0°C. Its height is estimated as 350 m. What pressure difference would result over a shaft of this height? If the shaft has an effective diameter of 3 m, what is the average velocity of air moving in the shaft? Assume it is fully open at the base and top of the building. (See Equations 9-3 and 9-4.) Let $f = 0.02$.
5. Assume a smoldering chair fire occurs in your classroom, and the doors and windows are shut. Using the information in Figure 9-16, estimate the time the smoldering fire smoke layer reaches the fire location at 1.2 m above the room floor (t_o). Also, estimate the time to achieve an incapacitation from CO at that location (t^*).

True or False

1. Fires in a room with a door or window open, producing smoke of less than 400°C, can be considered a developing, almost free-burning fire.
2. Smoldering fires typically produce lethal conditions in less than one-half hour.
3. Backdraft is a form of flashover.

4. In the summer, fire in a high-rise air-conditioned building can have smoke propagate to the floors below the fire.

5. A fire in a spaceship eventually has to go out due to oxygen depletion or it will burst the ship.

References

1. G. Hartzell, "Combustion Products and Their Effects on Life Safety," sec. 3, chap. 1 in *Fire Protection Handbook,* 17th ed., edited by A. E. Cote and J. L. Linville (Quincy, MA: National Fire Protection Association, 1991), 3.3–3.14.
2. J. G. Quintiere, K. Steckler, and D. Corley, "An Assessment of Fire Induced Flows in Compartments," *Fire Science and Technology* 4, no. 1 (1984): 1–14.
3. J. G. Quintiere and B. J. McCaffrey, *The Burning of Wood and Plastic Cribs in an Enclosure: Vol. I,* NBSIR 80-2054 (Gaithersburg, MD: National Bureau of Standards, September 1980), 49.
4. J. G. Quintiere, M. Birky, F. Macdonald, and G. Smith, "An Analysis of Smoldering Fires in Closed Compartments and Their Hazard Due to Carbon Monoxide," *Fire and Materials* 6, no. 3 and 4 (1982): 99–110.

Fire Analysis

Learning Objectives

Upon completion of this chapter, you should be able to:

- Find analytical applications in fire safety design and in fire investigation.
- Explain the benefits and limitations of fire modelling.

INTRODUCTION

Having completed the previous chapters, it should be clear that methods exist for estimating the size of a fire and its growth, $\dot{Q}(t)$ and for estimating its damage effects from heat and smoke. Also, it should be clear that this process is complex, and we have only considered simplified methods for estimating. Nevertheless, it is possible to step through a fire scenario using our equations with good results and reasonable accuracy. Most fire phenomena have uncertainties due to circumstances not completely known (e.g., the extent of a door opening, the proximity of one item to another). Even if we know all of these factors, our computations are not perfect, varying in accuracy by as much as $\pm$ 50%, but usually much better. Approximate analyses can still be very useful, even with these limitations.

In this chapter, we consider aspects of fire safety design and fire investigation. In fire safety design, a specific fire is not obvious; in fire investigation, we have some evidence or hypothesis of the fire. An important aspect of design is the specification of the relevant fire scenarios and their probabilities of occurrence. In investigation, we may use our calculations to clearly eliminate or assert a certain event or its time of occurrence. We discuss some examples and computer models that are often cited as the methodology for analysis.

FIRE SAFETY DESIGN

Design in fire safety is used principally in establishing compliance with the local regulations. These regulations, based on codes and standards of practice, specify the requirements. The architect, builder, or building owner must ensure compliance. To do more is not usually sought; however, many situations arise where the regulation is not clearly applicable or the designer wishes an equivalent alternative. The establishment of **equivalency** requires at least a technical discussion, but more appropriately an analysis. The formulas we have discussed for fire computations can be used for such analysis.

equivalency
the statement in regulations allowing for alternatives by design

Many aspects are subject to regulation and therefore open to design possibilities, including:

1. Detection and alarm,
2. Mitigation of growth and suppression,
3. Egress,
4. Continuity of operations,
5. Structural integrity,
6. Refuge and rescue.

We have not addressed all of these aspects in our discussion of fire behavior; they are subjects of other texts.

Perhaps the more significant aspect of fire safety design is the interaction of fire growth time with egress time. In general, we seek egress time to be less than fire damage time. For all possible fire scenarios and their likelihood (probability), we need to determine the egress time of the building population. Egress time increases as the fire progresses because smoke slows people. On the other hand, the time to cause damage by the fire is inversely related to fire size. As the fire grows, the time for damage decreases. Sprinklers, suppression by fire fighters, and fire resistance barriers all affect this damage time. A qualitative illustration is shown in Figure 10-1. Note that sprinklers can be defeated and the fire damage curve can turn downward again.

The two examples of fire safety design analysis have actually been conducted because of fire incidents; they were not conducted as part of the design process. However, these examples are introduced to demonstrate that such analysis is applicable in the design phase. The analysis of an actual fire incident demonstrates design strategies that could have been applied during development of the building, yet the building may have still been in compliance with the regulations after construction. Such analyses can provide cost-effective safety improvements or equivalency to the regulations.

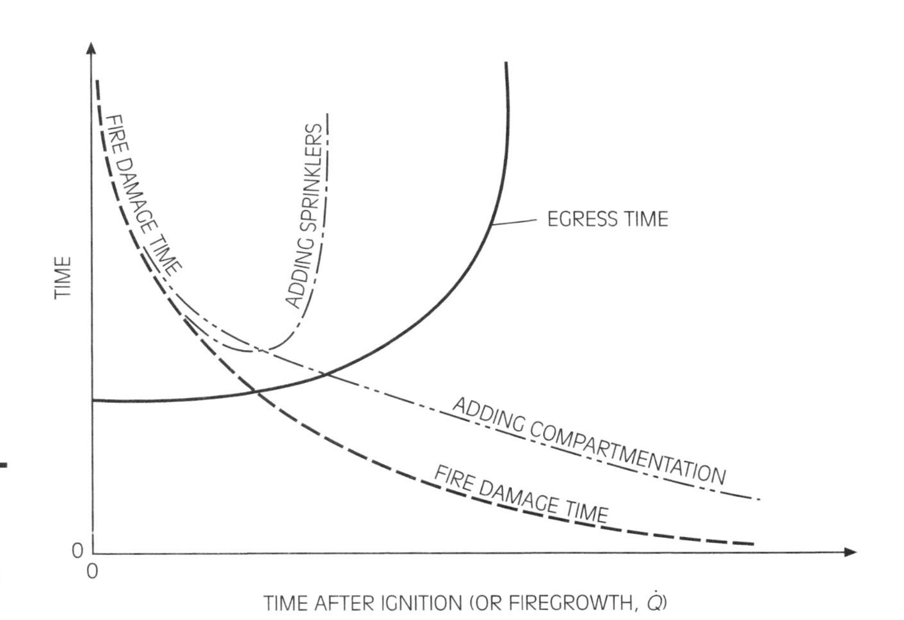

Figure 10-1 *Fire damage time compared to egress time for a given fire growth scenario.*

EXAMPLE 1

EFFECT OF SHAFT VENTS IN A BUILDING FIRE

The first example was computed by Takeyoshi Tanaka, a scientist of the Building Research Institute of Japan, using his computer code designed to calculate the smoke transport and its properties through a building.[1] The results of the computations are shown in Figure 10-2. Two shaft vent conditions are considered and lead to slightly different results for smoke movement and temperature. These computations were initiated following the MGM fire of November 21, 1980, in Las Vegas, Nevada. At the time computers could not easily or quickly compute the 65-story MGM building. So these calculations for only five stories were performed. Their inferences were never used for analyzing that fire or learning from its consequences. Most people (85) died on the upper floors of the MGM as smoke traveled up the vertical shafts. The calculations of Figure 10-2 show that the case of a vent in the center shaft does not affect smoke temperature very much, but does impact the smoke depth on the third floor. Further calculations could produce more complete information on the benefit of shaft vents.

EXAMPLE 2

SMOKE MOVEMENT IN THE WORLD TRADE CENTER, NEW YORK, NY

Another high-rise building fire attracted much attention, namely, the explosion and fire in the Word Trade Center in New York City in April 1993. The computer code illustrated in Figure 10-2 has since been updated (BRI II) and has been used to establish equivalent design alternatives in many Japanese building constructions. Computers now have much more capability and the entire 110 stories are included in the computations. These computations were implemented by J. Yamaguchi et al.[2] to analyze the fire scenario of the World Trade Center. An excerpt is shown in Figure 10-3. A fire of 20 MW for a 25 min was selected to simulate the parking garage fire (shaded black in Figure 10-3). The results show the smoke flow (black arrows for upper smoke layer flows, white for lower layer flows) at 13 minutes (780 s).

No direct deaths resulted due to the smoke filling the tower building, probably because of the duration of the fire and the ventilation conditions. Such calculations can not only estimate the smoke conditions under this fire event, but could be beneficial in assessing the hazard of other fire scenarios. Design strategies that might make the building safer can be evaluated on the computer.

PERFORMANCE CODES

performance fire safety codes regulations providing for engineering analysis

Today, there is much talk about **performance fire safety codes** as opposed to prescriptive codes. Essentially this means using computational analysis over set specifications in the regulations. This process is evolving and no fixed methodology has emerged. Because fire computations require certain expertise, performance codes will not easily be put into practice until the knowledge is fully exchanged

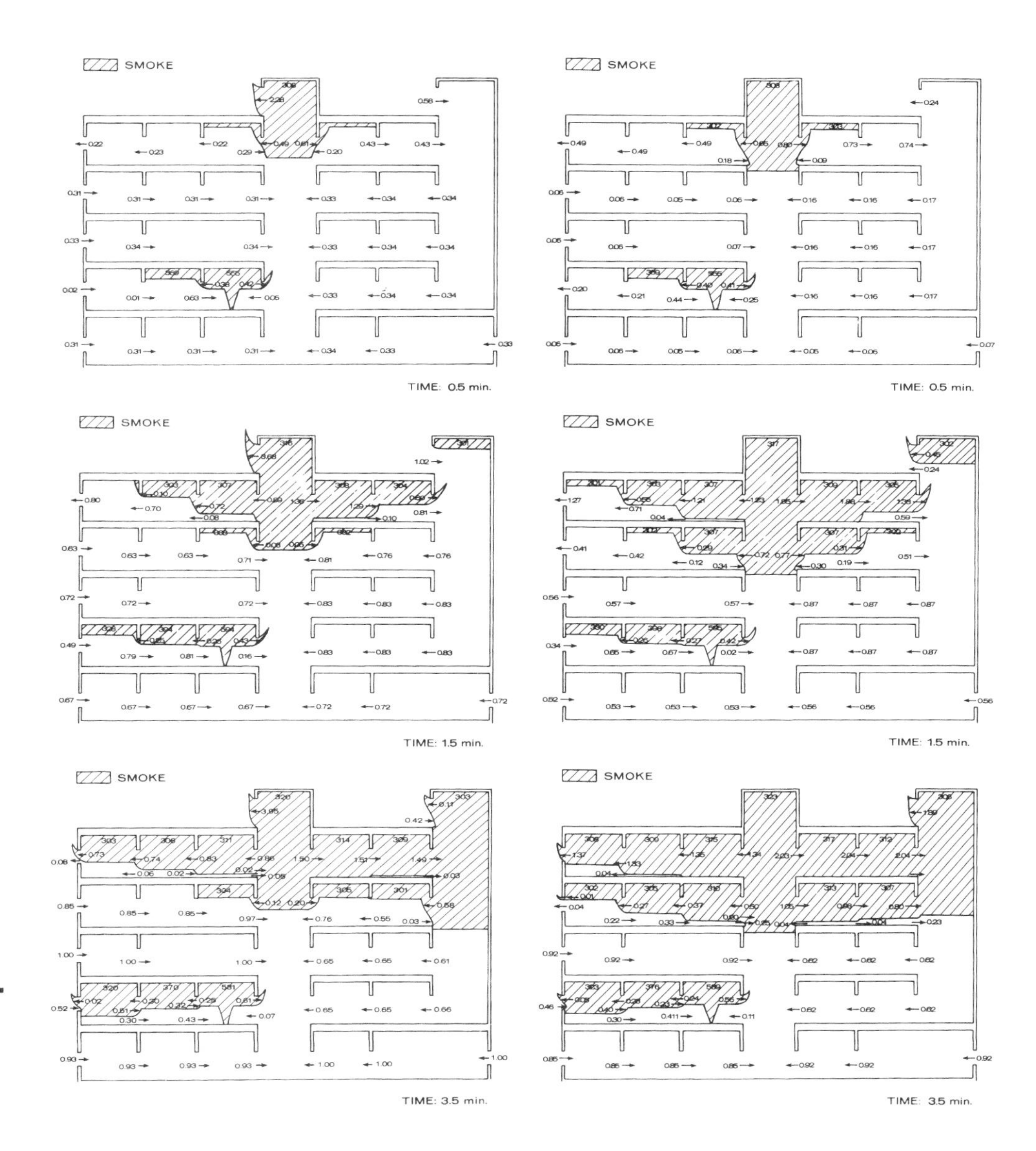

Figure 10-2 *Smoke movement computations. From Tanaka, Ref. 1.*

and appreciated at all levels of the process. Until this expertise is comprehensive throughout the practice, fire safety design will continue to address equivalency issues. Changing fire safety design from regulations to computers is technologically feasible but may not be practical to implement. Such new technology must progress hand-in-hand with needed research to fill the knowledge base and the computer gaps.

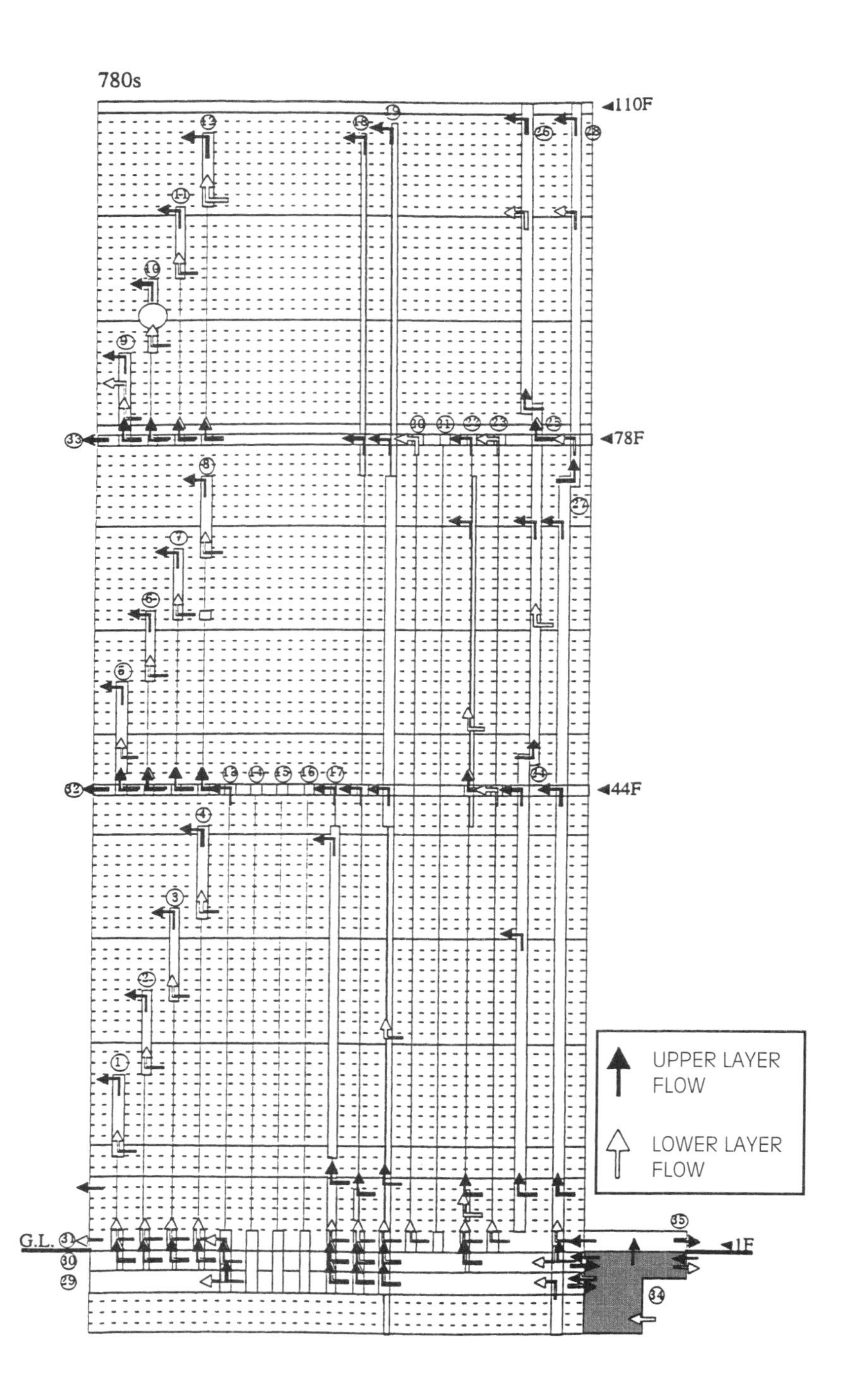

Figure 10-3 *Analysis of smoke movement in the World Trade Center fire, New York City. From Yamaguchi, Ref. 2.*

FIRE INVESTIGATION EXAMPLES

Litigation has pushed to use of computers and fire analysis into the field of fire investigation. As in design applications, there is no simple approach without a competent knowledge of fire behavior. Although computer models can be useful, the knowledge gained in this book can go a long way to bolster a timeline or proposed scenario in a fire investigation. Unlike design, this process for investigation is not open ended. There was a specific fire and it did specific damage. These facts must be established through evidence, witness accounts, and analysis. Let us examine some specific examples that illustrate the process of analysis.

EXAMPLE 3

THE CASE OF THE LAUNDRY BASKET FIRE

In May 1993, the murder of John Martorano led to the conviction of his wife with an additional conviction of arson. A conspiracy was hatched by Donna Martorano with Jimmie Lee McWhirter to torch her residence while the family was in Florida. The following news articles detail the background:

Arson, fraud case going to trial

By Pamela Sampson

TRIBUNE-REVIEW

An Allegheny County woman who is a suspect in the murder of her husband is in the Allegheny County Jail on $400,000 bond, awaiting trial on charges that she tried to have her house burned down to collect insurance money.

Donna Martorano, 42, of 518 Rothey Drive in Elizabeth was held for trial on charges of arson, fraud and two counts of criminal conspiracy after a preliminary hearing yesterday before District Justice Jules Melograne.

She is accused of masterminding a botched attempt to burn down her home on April 19, and of filing a false theft report so insurance money could be collected for items including jewelry and clothes.

Testimony to back those accusations came from Martorano's stepdaughter, 13-year-old Lisa Martorano. She is the daughter of the late John Martorano, who was found shot to death behind his home on May 13.

Never losing her composure and intently avoiding eye contact with her stepmother, the girl testified that Donna Martorano had devised a scheme to have their house set ablaze while the family went to visit Martorano's son in Florida.

The stepdaughter said the defendant arranged for Jimmie Lee McWhirter of Texas to come to Pittsburgh and torch the residence while the family was in Florida. McWhirter is Martorano's cousin; according to court documents, the two were raised practically as brother and sister and have a close relationship.

To save some valuables before the fire, according to the girl, the family piled items including clothes, a videocassette recorder, a computer and a television into their van and then left the state.

But the alleged plan didn't work out. According to testimony from a county fire official, investigators believe a fire was deliberately set in the basement, but a water pipe quickly melted and the water extinguished the blaze in less than 30 minutes.

The stepdaughter testified that on the day of the fire, Donna Martorano called her after and was told the basement was filled with smoke and water, but the house was still standing.

According to the girl, her stepmother was angry and hung up the phone, stating

(*continues*)

(continued)
"Jimmie Lee screwed up; he was supposed to burn the house down."

On the trip back to Pennsylvania, the girl testified, Donna Martorano decided to report that a theft had occurred at the house while the family was away, and told each family member to report certain items of clothing, jewelry and money—as missing.

Under questioning by Assistant Public defender Sumner Parker, the girl admitted she and her late father willingly participated in the plans even though she knew they were illegal.

Although the defendant faces serious charges, they do not at the moment include homicide. She was arrested last month and charged with plotting the murder of her husband, but prosecutors withdrew the charges earlier this week. They may be refiled at a later date, according to Deputy District Attorney W. Christopher Conrad.

In the meantime, McWhirter and an alleged accomplice, Marlin Dale Collins of Texas, remain accused of killing John Martorano.

According to an affidavit of probable cause filed by county homicide detectives, police believe Donna Martorano arranged to have her husband killed after she found out he was going to take his daughter and leave.

Witnesses told police that John Martorano feared for his life because his wife had threatened to kill him if he did not cooperate with her plans to collect the insurance money, the affidavit states.

According to the affidavit, McWhirter told police that his cousin "solicited him to hire Marlin Collins" for the hit and offered to pay Collins $5,000.

Courtesy of the *Tribune Review*, Pittsburgh, PA

Wife guilty in hired killing

Defense pleads for life in prison

By Jan Ackerman
Post-Gazette Staff Writer

A defense attorney last night begged a jury to give an Elizabeth Township woman life in prison—not death—for hiring two men to kill her husband as he jogged in Jefferson last May 13.

"If Donna Martorano had the guts to pull the trigger herself, you would not be sitting here today," assistant public defender Robert Foreman told the jury during the death-penalty phase of Martorano's trial, which began at about 8 p.m.

He was referring to the fact that a contract killing is a basis for the death penalty, but first-degree murder without such aggravating circumstances does not automatically carry a death sentence.

Yesterday afternoon, the jury found Martorano guilty of first-degree murder, criminal conspiracy, two counts of solicitation to commit murder, arson, conspiracy to commit arson, insurance fraud and conspiracy to commit insurance fraud. Jurors deliberated for three hours and 15 minutes.

Deputy District Attorney W. Christopher Conrad argued that Martorano deserved the death penalty because she tried to pay another person to kill her husband of three years, John Martorano.

Foreman asked the jury to consider as a mitigating factor that Martorano had no prior criminal record, had been a good mother, and had worked at various jobs to support her family.

Her 18-year-old son, Jason Gulyban took the witness stand and told the jury that he loved his mother and didn't want her to die.

Then Foreman made an emotional plea on behalf of Martorano, telling jurors that no matter what their decision, she would never see another movie, go to the theater, a ball game or a restaurant.

"She will never be able to attend her children's weddings or hold a grandchild at a christening," Foreman said. "Her punishment will be sufficient. You do not have to kill her."

Common Pleas Judge Donna Jo McDaniel sent the jury of six men and six women out to deliberate about 8:45 p.m., but within 20 minutes, jurors said they were tired and wanted to retire for the night. They were housed in a Downtown hotel, and deliberations were to resume this morning.

The first-degree murder conviction pleased members of the Martorano family. John Martorano's mother, Rose, and his daughter, Lisa, 14, both hugged Conrad.

"I am just happy that she got what she deserved," said Lisa, who testified against her stepmother.

Conrad based his case on the theory that Donna Martorano contracted to have her husband killed because he was about to go to authorities and report a bungled arson of their home.

Last March, the Martorano family was notified that a bank wanted to foreclose on their split-level house on Rothey Drive in Elizabeth Township.

Conrad said Donna Martorano reacted to that news by nearly doubling the insurance on the house and contacting her first-cousin, Jimmy Lee McWhirter, 55, of Fort Worth, Texas. Conrad said she hired McWhirter to burn the house down while the family was in Florida for Easter.

When the house caught fire, a water pipe burst and extinguished the flames. McWhirter told the jury he set up the arson, but did not actually burn the house.

Conrad said Martorano submitted false claims to the insurance company for electronic equipment, jewelry and cash she claimed was stolen from the house at the time of the fire.

Conrad used three types of evidence to prove that Martorano arranged the contract murder of her husband on May 13: the direct testimony of McWhirter and Marland "Doc" Collins, 34, the two men who said they killed her husband; admissions that McWhirter and Collins made to friends and relatives in Texas; and telephone calls between Martorano and McWhirter.

Conrad apologized to the Martorano family and to the jury for making deals with McWhirter and Collins that ensured they would not get life sentences or the death penalty. Both men pleaded guilty to third-degree murder as part of a deal to testify against Martorano.

Courtesy of the *Post-Gazette*, Pittsburgh, PA

One issue in this case was the fire in a laundry basket that ruptured a copper water pipe, extinguishing the fire. The evidence showed that the fire had been intentionally set between the hot water heater and furnace in the laundry room of the residence. An overhead copper water pipe opened at an elbow joint as the solder connection melted. Plumbing solders melt in the range of 250° to 300°C at most. To achieve such a temperature, the flame tip of the fire must reach the overhead copper pipe, as can be seen from Figure 7-10. The ignition source must be, at least, capable of this temperature at this height. Moreover, to melt the solder in a water-filled pipe requires the pipe to be bathed in the flame to maximize heat transfer to the solder connection. Figure 10-4 illustrates the scenario. An ordinary laundry basket fire could be approximated as a large wastebasket fire from Table 6-4 as 10 g/s at an estimated heat of combustion of 20 kJ/g (assuming plastic and cotton clothing). This gives

$$\dot{Q} = (10 \text{ g/s})\,(20 \text{ kJ/g}) = 200 \text{ kW}.$$

From Equation 7-9 we compute a flame height of

$$\begin{aligned} L_f &= 0.23\,\dot{Q}^{2/5} - 1.02D \\ &= 0.23(200 \text{ kW})^{2/5} - 1.02(0.7 \text{ m}) \\ &= 1.2 \text{ m} \end{aligned}$$

where we have assumed a diameter of 0.7 m. This flame height, 1.2 m, represents probably an upper limit for a laundry basket fire. But we see that the flames do not reach the pipe at the ceiling (a). On the other hand, if an accelerant—say gasoline—were added to the laundry basket, then we have (from Chapter 6):

$$\dot{Q} = \left(55 \text{ g/m}^2 - \text{s}\right)\left(\frac{\pi}{4}(0.7 \text{ m})^2\right)(43.7 \text{ kJ/g}) \text{ for gasoline}$$

or

$$\dot{Q} = 925 \text{ kW}$$

and

$$L_f = 0.23(925)^{2/5} - 1.02(0.7)$$

or

$$L_f = 2.82 \text{ m}$$

This flame would strike the ceiling and easily bathe the copper pipe joint. Its temperatures would be sufficient to melt the solder (b).

It should be clear from this example that some choices had to be made on the nature and size of these fires. The assumptions could have been different, but the conclusions would not be easily changed. An ordinary laundry basket, even if somehow accidentally ignited, is not very likely to be capable of melting the solder. Even without any direct evidence of an accelerant, such an analysis would strongly indicate its presence.

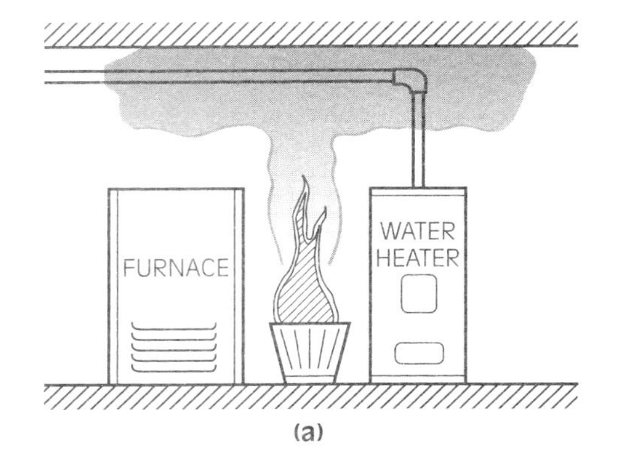

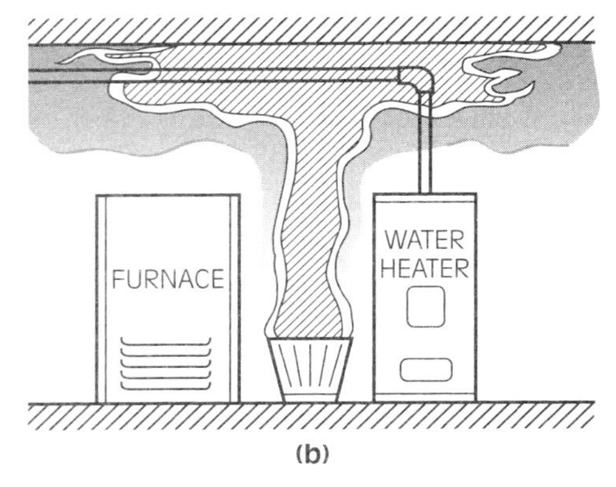

Figure 10-4 *The laundry basket fire and the copper pipe joint.*

The foregoing example arose in a training course given to fire investigators in the Bureau of Alcohol, Tobacco, and Firearms (BATF) program. Agent William J. Petraitis posed the problem after he and Allegheny County investigator Thomas Hitchings had conducted tests on laundry baskets with and without accelerants. They were looking for tell-tale fire signatures that might occur. The flame height is the key indictor here. The example proved useful to illustrate the application of the science in this course to a real case.

EXAMPLE 4

AN ANALYSIS OF THE WALDBAUM FIRE, BROOKLYN, NEW YORK, AUGUST 3, 1978

On August 3, 1978, a fire occurred at the Waldbaum supermarket in Brooklyn, New York. A sketch of the store is shown in Figure 10-5. At the time of the fire, an extension was under construction as indicated. The store was a typical supermarket with a mezzanine along a portion of the north wall. The loft was completely of wood construction comprising the floor, roof, and structural support trusses. The Trusses were made of 3 by 12 in. members interlaced together in bundles of 4 or 5. Trusses 2, 4, and 6 were covered on one side with plaster to form fire walls in the loft. However, to enable passage, these trusses had doorway openings. The roof had been modified with an added rain roof at the peak, which formed a double layer roof at the peak. Also the new construction required a splayed roof section to meet the new roof of the extension, forming another double roof triangular section along the north wall as shown in Figure 10-6. Due to the construction and the addition of new columns along the wall to the extension, construction voids existed along this wall.

Construction work began at 7 A.M. and the store opened for business at 8 A.M. Flames were first seen at 8:30 A.M. along the interface between the ceiling and extension wall of the mezzanine men's room and the machine room. The fire eventually spread into the loft between truss sections 4 and 6, and collapsed truss 5 at approximately 9:15 A.M., causing twelve firefighters to fall into the flames. Six were killed.

A man was tried and convicted in 1978 of setting this fire. His confession stated that he and two others set the fire near dawn (approximately 6 A.M.) by making holes in the roof and using newspaper and lighter fluid to initiate the fire below. This confession was later questioned and discounted in a retrial that was held in 1994.

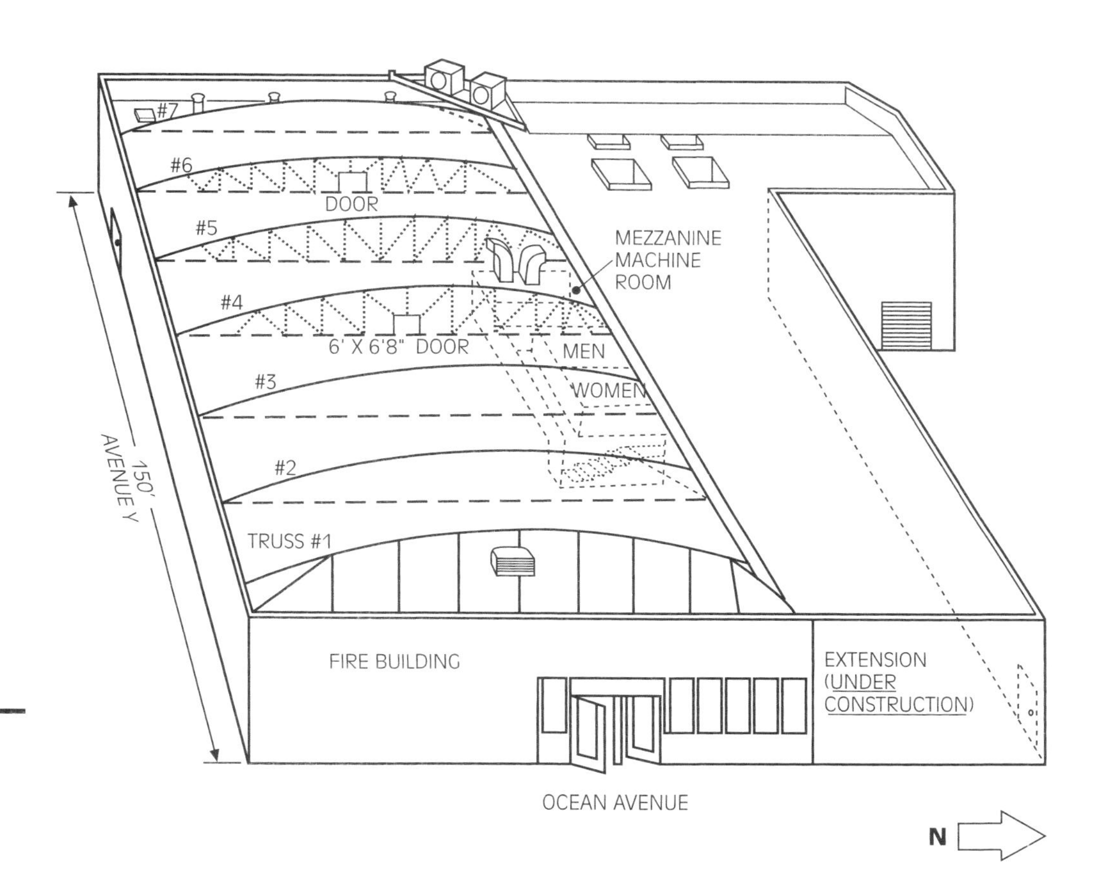

Figure 10-5 *Waldbaum Supermarket, Brooklyn, New York, 1978.*

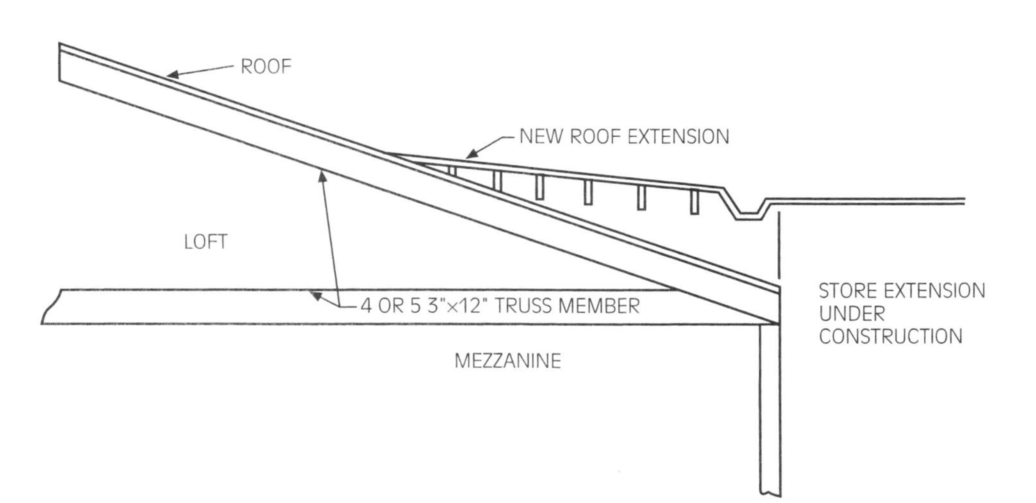

Figure 10-6 *Splayed roof extension and structural support trusses. The trusses were made of 3 × 12 in. members. Five members were interlaced.*

A consistent fire scenario was never fully presented. The original fire investigators could not agree on a cause and later suggested that the cause of the fire may have been of electrical origin. However, there was no electrical power in the ceiling area where flames were first seen, and a man in the loft at the onset of the flames saw no evidence of fire. The prosecutors sought advice on how to explain an alleged fire starting at 6 A.M., but not seen until 8:30 A.M.

I was asked to assist William J. Petraitis, special agent of the Bureau of Alcohol, Tobacco, and Firearms in the investigation of this fire 16 years after it occurred. Agent Petraitis discovered the splayed roof extension, shown in Figure 10-6, and reasoned the fire had to begin in that space. We then examined the hypothesis of the fire beginning there at approximately 6 A.M. to see if it could be consistent with the other known events. A fire scenario was developed and calculations were examined to support the plausibility of the events and their duration. All of this analysis is not presented here, but the results are described.

EARLY DAWN: IGNITION (APPROXIMATELY 6 A.M.)

At approximately 6 A.M., an intentional fire is considered to have been set in the roof area adjacent to the construction of the building extension (Figure 10-7). The fire is set by stuffing paper through holes in the new roof extension along with a liquid accelerant. The splayed roof extension has been built over the existing roof and forms a void space between the two roofs. The new roof is supported by rafters. The fire is set in channels of the wood rafters that extend between the primary wood trusses 4 and 5. Gaps under the rafters allow air to flow into the fire, but the space is mainly enclosed with temporary partitions at the wall adjacent to the new building extension. The accelerant-soaked paper probably caused a fire of 100 to 500 kW in one or more rafter channels, involving no more than 1 m^2 (Figure 10-8).

Under expected heat fluxes of 40–50 kW/m^2, the wood members would ignite in 30 s to 1 min, and begin to contribute roughly 500 kW. The accelerant fire would

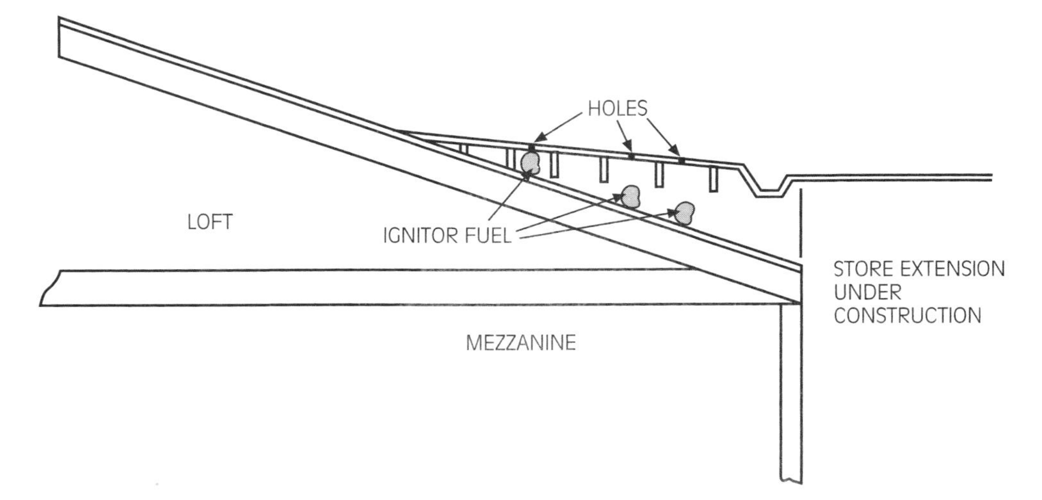

Figure 10-7 *Flammable-liquid-soaked newspapers inserted into roof holes.*

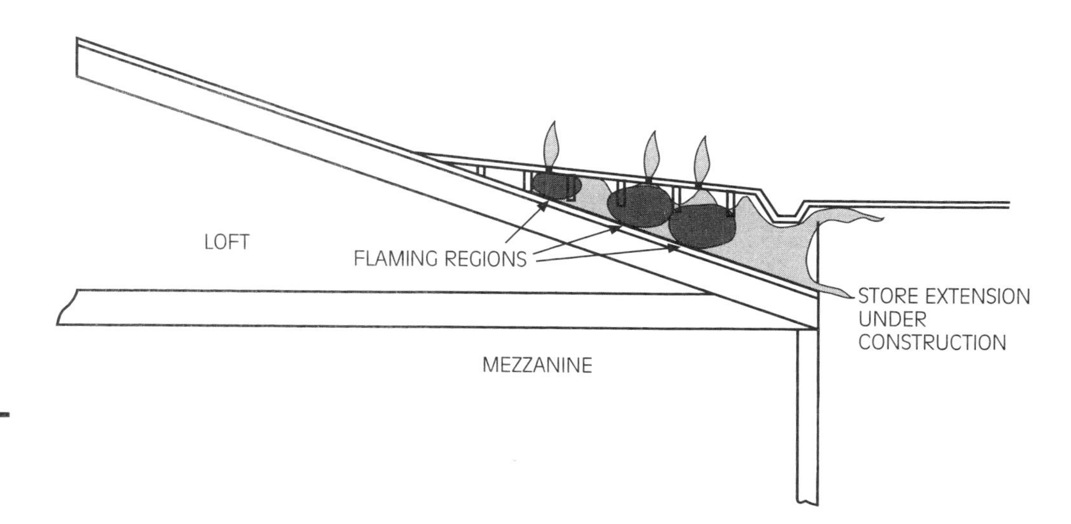

Figure 10-8 *Ignition at approximately 6 A.M.*

probably burn out in 1 to 2 min. The wood fire could progress to roughly 1000 kW, but then it would become limited in further growth due to combustion products filling this confined space.

SMOLDERING STAGE

Oxygen depletion would cause flaming to cease in about 1–2 min following wood ignition. However, sustained smoldering would be possible, especially in the smaller rafter spaces where radiant heat transfer would be high. See Figure 10-9. Ohlemiller[3] reports that sufficient radiant heat transfer is necessary to sustain smoldering in wood cavities. He also reports smoldering propagation rates of 1 to 6 cm/hr. Using the smaller value, smoldering can propagate through the old 3/4-in. plank roof in roughly 2 h. A short time before 8:30, a hole would form between the loft and the roof cavity. With air velocity increased to 25 cm/s Ohlemiller[3] found smoldering in wood would change to flaming combustion. Such velocities would be realized at the hole as air was naturally induced to flow from the loft into the hole. Flames would erupt in the false roof cavity space.

ONSET OF FLAMING: SHORTLY BEFORE 8:30 A.M.

At approximately 8:30 A.M. flames are observed in the mezzanine area at the wall and ceiling near truss 4. It is believed that this was due to the expansion of the flames as flashover caused a rapid rise in energy release rate in this confined space (Figure 10-10). The associated pressure increase forced flames through available void spaces. As the hole to the loft became larger, the increased air flow rate would also promote a larger flaming fire in the rafter spaces. The flow of the flaming combustion products would follow the rafter channel to the end points at trusses 4 and 5. The first flames are seen at an opening near truss 4. As the hole to the loft became even larger, flames would lap upward under the sloping wood ceiling of the loft. The person in the loft before this time would not have been aware of the fire because smoke would not permeate into the loft until the hole became large. In fact, he only saw flames when he descended to the machine room after being warned of the fire.

Evidence to support this scenario is shown in the photograph of Figure 10-11. The small white vapor

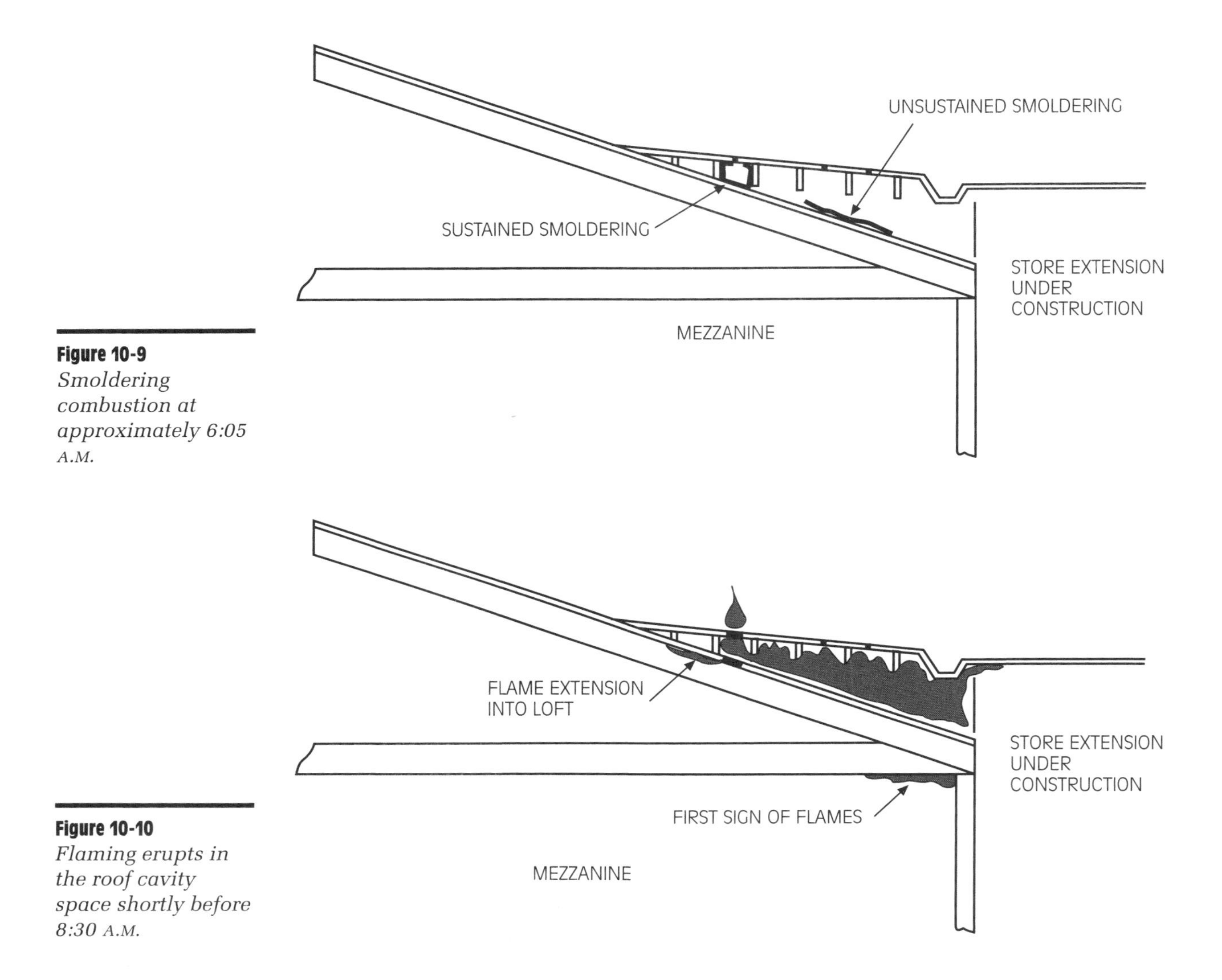

Figure 10-9 *Smoldering combustion at approximately 6:05 A.M.*

Figure 10-10 *Flaming erupts in the roof cavity space shortly before 8:30 A.M.*

plume rising from the roof indicates a hole in the roof. This hole is unlike the larger roof holes, attributed to the fire fighting, on the fire-wall protected side of truss 4 where grey smoke is emanating. Flames are seen to have emerged from the roof on the other side of truss 4. This region is likely just above the first and highest hole over the most confined rafter space as shown in Figure 10-7. Smoldering would have been sustained in this area. The hole indicated by the white vapor plume is in the area of the larger void space where fire was likely initiated, but smoldering was not sustained.

FIRE GROWTH IN THE COCK LOFT

Flame spread would rapidly move from the hole, up and under the loft wood ceiling. It is estimated that

Figure 10-11 *Firefighters on roof of Waldbaum store before roof collapse. Courtesy of B. J. Hughes, Ref. 5.*

flames would move from the hole region up and along the peak of the loft to truss 4 and through truss 5 in 2–3 min. This time was estimated from a flame spread model[4] using an ignition time of 30 s and an energy release rate per unit area of 150 kW/m^2. Such calculations are beyond the information presented in Chapter 5, but are based on the theories explained there. This rapid spread (Figure 10-12) seemed incredulous. With the low confidence in flame spread calculations, an experiment was performed in a similar, but smaller loft. Flashover occurred in the experiment in approximately 1 min. In the store loft, flashover was estimated to occur when the wood fire contribution was 6 to 12 MW in the section of the loft between trusses 4 and 6. The flame spread calculation indicated that this would occur in approximately 2 1/2 min. Full involvement was estimated to take up to 5 min more. From the onset of flaming at the hole, full involvement of the loft section would take approximately 5 to 10 min. Because of the false roof sections and the truss fire-wall sections, the firefighters on the roof were unaware of the raging fire in the loft section below them.

COLLAPSE OF THE ROOF DUE TO TRUSS MEMBER FAILURE

If one assumes that the roof collapse is due to the failure of a truss member as a result of fire degradation, the burn-through time of a member can be estimated. The fire exposure assumed is a truly involved loft with a 3-in. thick truss member within the flame. Using a mass burning rate per unit area of 11 g/m^2 s, typical for wood under these conditions, leads to a burn-through time compared to the recorded roof collapse 5 min later.

CONCLUDING REMARKS

This fire scenario was a complex series of phenomena that are not usually appreciated in the study of fire.

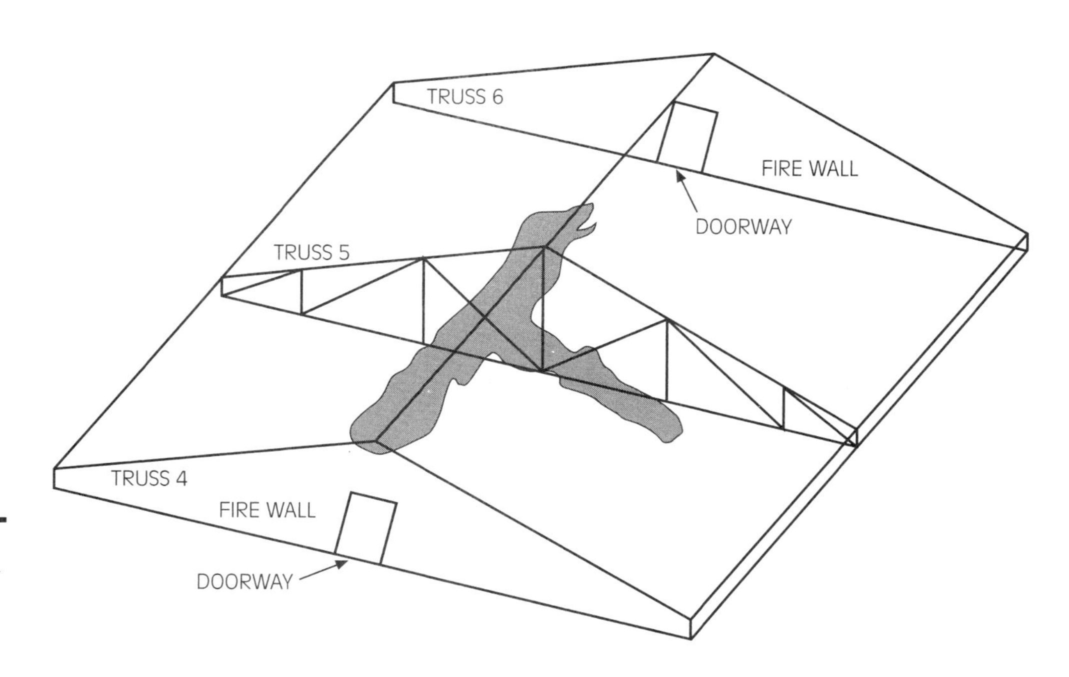

Figure 10-12 *Fire growth in the loft at approximately 8:35 A.M.*

Experienced fire investigators were not able to deduce this process. Yet relatively simple scientific analysis could produce a plausible and consistent series of events within the periods of the recorded observations of this fire. Some may question the details, but the scenario fits the timeline of what is known and alleged. The photograph in Figure 10-11, which shows the small white smoke plume in the lower right below the bottom edge of the flames is a tell-tale indicator. It would appear to be one of the original ignition holes where smoldering was not sustained in the splayed roof structure. But for a hole above, the fire went from flaming to smoldering and back to flaming. The low rate of downward flame spread kept the fire in the vicinity of the upper hole; but the more rapid upward spread drove the fire into the loft with tremendous speed.

This scenario can not be described or defended without a comprehensive knowledge of fire behavior and available data to support the calculations. Smoldering research and data are rare but were critically needed in this analysis. The background of this fire was provided by the Office of the District Attorney, Kings County, Brooklyn, New York.[5]

EXAMPLE 5

THE BRANCH DAVIDIAN FIRE NEAR WACO, TEXAS (APRIL 19, 1993)

It is rare that data are recorded during a fire where investigators have a visual documentation of what took place. Because of the circumstances of the siege at the Branch Davidian compound near Waco, Texas, in 1993, much video and photographic recording equipment was being used. The news media were recording video approximately 1 to 2 mi. from the compound, and the government authorities had overhead surveillance aircraft shooting still photographs and a precise time-coded near-infrared video (FLIR) camera. Because of the thermal sensitivity of the FLIR, it easily recorded the fire inception. From these video data, the information in this text, and the behavior of fire, it was possible to reconstruct the characteristics of this fire. A summary of the results were contained in testimony before the Congressional Subcommittee on Crime of the Committee on the Judiciary, and the Subcommittee on National Security, International Affairs, and Criminal Justice of the Committee on Government Reform and Oversight.

CONGRESSIONAL COMMITTEE STATEMENT ON THE MOUNT CARMEL BRANCH DAVIDIAN FIRE, APRIL 19, 1993[6]

Background

My name is James Quintiere. I am a professor of Fire Protection Engineering at the University of Maryland, College Park, MD. Before coming to the University, I was a Division Chief in charge of fire research at the Center for Fire Research of the National Institute of Standards and Technology. I have 25 years of experience in fire research, education, and in the science of fire growth. I am currently chairman of the International Association for Fire Safety Science, a world organization of scientists and engineers for the promotion of fire research and its beneficial applications.

Shortly after the fire of the Branch Davidian compound at Waco, Texas, on April 19, 1993, I was asked to contribute to the fire investigation. In doing so, I enlisted the support of Dr. Fred Mowrer, also of the Department of Fire Protection Engineering, University of Maryland. We visited the Waco fire site during April 22, 1993. At that time, we joined the team under Paul Gray (Houston), which also consisted of Thomas Hitchings (Pittsburgh), William Cass (Los Angeles), and John Ricketts (San Francisco). The group under Paul Gray would focus on the cause and origin of the fire. We would analyze the development of the fire and draw interpretations and conclusions from that analysis.

Visual Data

The fire had completely leveled the compound so that no significant structural remains were available to establish the development of this fire. However, this fire was probably one of the most extensively recorded fires in history. Not only were commercial television stations continuously recording this event, but surveillance government planes were taking still photographs and using a forward-looking infrared (FLIR) video. These visual records became the principal source of data for our analysis.

The video and photographic data were made available to us by the FBI. Video copies of data we requested were given to us at the FBI Headquarters in Washington, D.C., on April 25th. Subsequently, the FBI video and photo laboratories supplied additional materials and support as requested during our investigation. The data included television coverage of the fire by the Canadian Broadcast Corporation (CBC), by Channel 10 of Waco, the FLIR video recording, and aerial photographs. These covered the time period of the fire, approximately 12:00 to 12:30 P.M. CDT.

The principal source of data to establish the inception of the fires and their locations is the FLIR video. Based on the calibrated clock of the FLIR, the other video and photographic records could be correlated, and a comprehensive visual record of this fire could be established. From this visual data, I was able to determine the point of origin of the fires, the growth rates, and estimates of the fire energy output rates at critical transition points in their development. I also drew conclusions of the nature of the ignition sources, the role of the tear gas, the effect of the wind, and the survivability time of the occupants. I will summarize these conclusions and how they were determined. In addition to this statement, I would like our official report and a video I made for the criminal trial to be submitted for the record of this hearing. If you wish, I can review the video as well.

Ignitions

At least three separate fires began in the compound on the day April 19, 1993.

Fire 1: The first began at 12:07:42 P.M. CDT in the front of the second floor right tower. This is believed to have been a bedroom. We can expect the furnishings to be indicative of a crowed bedroom. I counted about 7 mattress box springs remains in the fire debris at this general location, presumably from this and adjoining rooms.

The precise time of the onset of this fire can be determined because of the characteristics of the FLIR camera. The FLIR camera records the intensity of light and heat radiation in the wavelength range of 8 to 12 micrometers (μm). This is in contrast to what our eye sees, which is in the range of 0.4 to 0.7 μm. As a result, the FLIR operated on autoranging, which would set the center of its black and white shading, or gray scale to the ground temperature (say roughly, 81°F). Then it set its range 40° above and below this midtemperature. As an object in the field of view emitted more radiation due to a temperature increase, the object would appear more white on the IR video. For an 81°F midrange temperature, this would mean that a change from gray to white color would indicate a temperature increase to 120°F or higher. Reflected sunlight could also cause white images, and the FLIR could penetrate smoke; but as the smoke became hotter and thicker, it would see it as white smoke. The FLIR sensor would become saturated at 194°F, above which the image would not be distinguishable. Consequently, the FLIR could detect, by a color change to white, temperatures as low as approximately 160°F (±30° due to autoranging). And, it could see through much of the early light smoke of the fire that would obscure the building by normal viewing. The FLIR is the definitive key to the detection of these fires.

The image of the temperature rise of the first fire is seen in Figure 10-13, in the second floor south corner bedroom. The first sign of this temperature rise was seen at 12:07:42 in the front side of this room. The image in this photograph occurred 9 seconds later, and is due to the transport of hot gases within the room.

In a similar manner, the other fire starts were determined. It appears that they all began on the perimeter of the building.

Fire 2: The second fire began in the dining room

Figure 10-13 *FLIR image 9 seconds after fire seen in second floor south bedroom. A white heat image is seen at the corner window.*

on the first floor level, approximately one minute after the first fire, at 12:08:48 P.M. CDT. This is seen on the FLIR video by a hot plume rising from the rear of the dining room. On surveying the fire debris, I counted 20 burned stacked chairs in this general location within the dining room.

Fire 3: Nearly, one minute after the dining room fire began, the third fire is seen in the chapel window on the right side of the building at 12:09:45 P.M. CDT. This is shown in Figure 10-14. The dining room fire is also visible, and the bedroom fire has now affected adjoining rooms, adjacent and above.

Less than a minute later, a related or separate fire is seen to occur in the debris area behind the chapel at

Figure 10-14 *FLIR image showing hot (white) region at first floor chapel window indicating start of third fire.*

12:10:23 P.M. CDT, as shown in Figure 10-15. This fire could have been connected to the previous chapel fire. The time difference between the two fire observations is comparable to the time associated with flame spread on a liquid fuel poured between the two points.

Figure 10-16 shows an aerial view at about the time of the possible fourth fire start. By comparing this figure to the previous figure, it can be seen that the visible smoke is much more evident than in the FLIR image of Figure 10-15. This figure shows the advantage of the FLIR in being able to see through this smoke.

Flashover

Following ignition of these fires, the next significant event is flashover, which marks the transition point of a

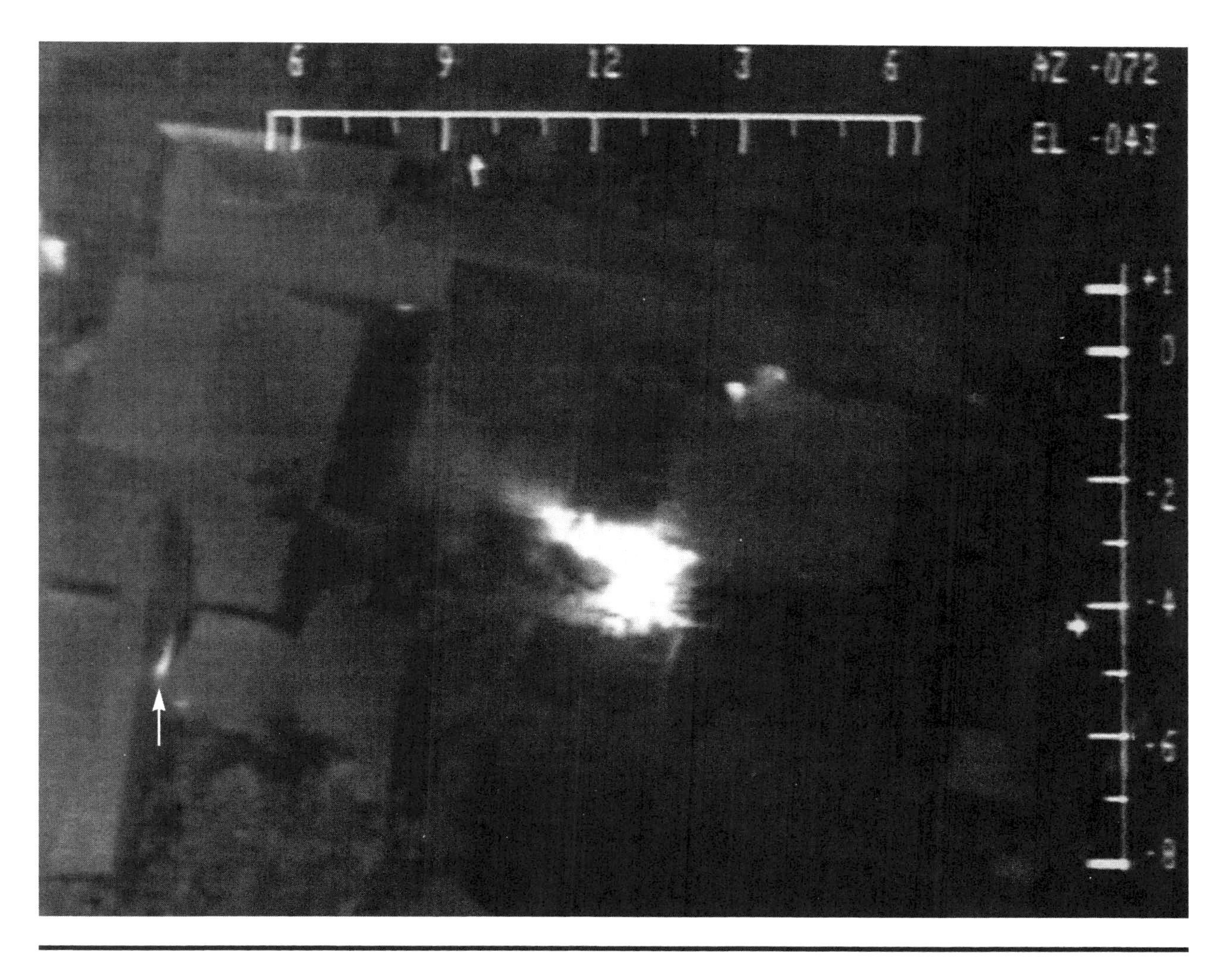

Figure 10-15 *FLIR image showing possible 4th fire start in debris area behind chapel.*

discrete fire in a room to a fully developed fire in which flames now fill the room and emerge from the windows. It is rapid and can take place in seconds. It occurs after the room is sufficiently heated. It marks the difference between survivable and nonsurvivable conditions in that room. These events can be seen, directly and indirectly, from the video records. The first is seen directly for Fire 1 as shown as window flames appear in the split screen images of Figure 10-17.

Flashover occurs at 12:09:42, two minutes after the start of that fire. Calculations show that this fire growth rate for the initial burning item would be rated as "fast" according to NFPA standard 72E. Its energy release rate would be about 2 megawatts (MW) at flashover, compared to an estimated 50 kilowatts (kW) that was necessary for detection by the FLIR. The detectable fire is like a 1-ft^2 spill of gasoline, compared to a 10-ft^2 gasoline fire at flashover.

Fires 2 and 3, in larger rooms, grow much more rapidly than the bedroom fire. Flashover occurs in about 2.5 minutes for the dining room (12:11:07), and in 4 minutes for the chapel (12:13:49). Figure 10-18 shows

Figure 10-16 *Normal photograph showing smoke obscuration of the compound.*

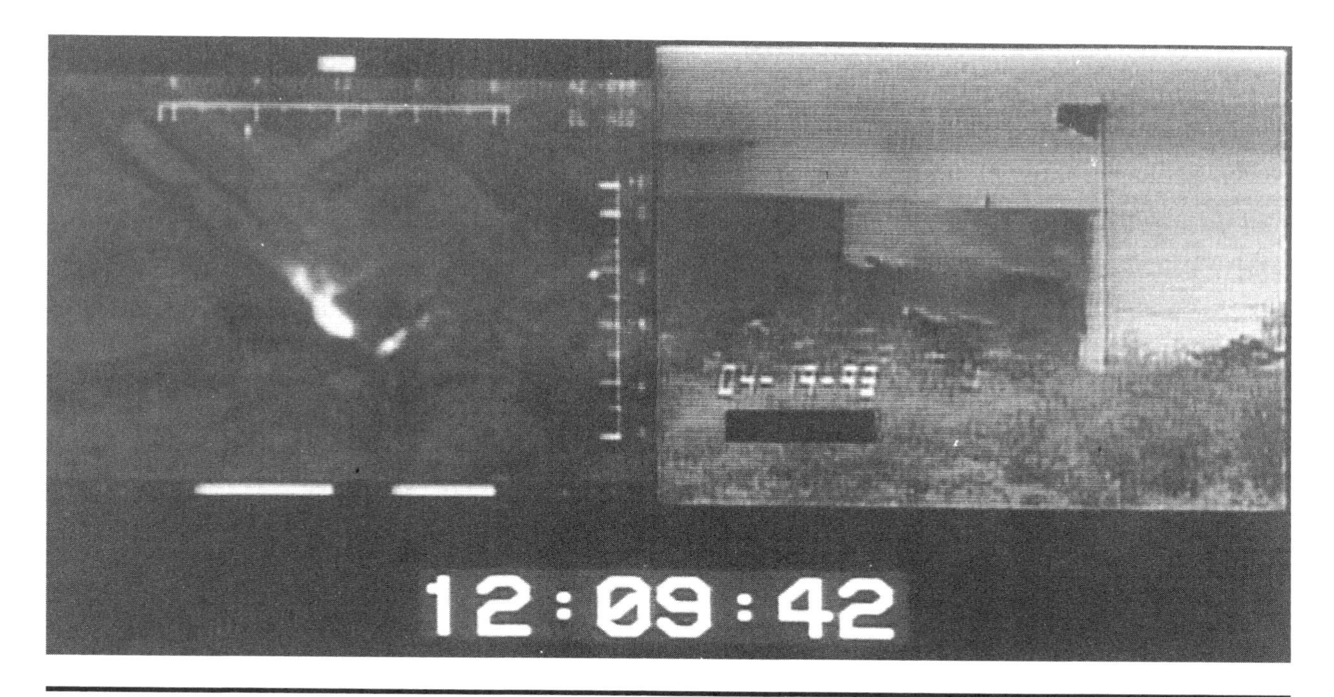

Figure 10-17 *Split screen showing a comparison between an FLIR image and a normal video image at 12:09:42.*

the effect of flashover for the chapel by black smoke that suddenly emerges from the front opening in the building. This smoke, pouring into the 25-mph wind, is due to the overpressure caused by the sudden increase of energy associated with flashover in the chapel.

Fire Cause

It is concluded that these three fires, occurring nearly at 1-min intervals, were intentionally set from within the compound. Even if the tank battering had caused the spillage of fuel from lamps, a match would be needed to initiate the fire. An electrical spark is ruled out because the electric power was shut off in the compound. It is obvious that these three fires needed an ignition source deliberately placed in each of the three locations. Also, none of these three fires could have caused any of the others because their growth rates would not provide sufficient heating to cause such remote ignitions. Any external heat source that might have been used to start the fires would have clearly been visible on the infrared video. This heat source was not seen. Although normal furnishings and interior construction characteristics would provide a means for fire propagation, the more than usual rapid spread of these fires, especially in the dining room and chapel areas, indicates that some form of accelerant was very likely used.

Tear Gas

Methylene chloride, used as a dispersal agent for CS tear gas, is flammable as a vapor at a concentration of 12% in air; however, it is not easily ignited as a liquid. In fact, it will put out a match on attempting to ignite the liquid. Although fire spread was relatively rapid in the compound, these rates are not indicative of the much more rapid propagation that would be associated with

Figure 10-18 *Black smoke is pushed out of the front (right) opening as a result of flashover in the chapel.*

a flammable mixture in air. Those rates would be in excess of 2 ft/s, and would be seen as a fireball moving through the atmosphere of the interior of the compound. No such characteristics were observed in this early fire growth.

Recently, I conducted additional experiments to access the role of methylene chloride as a wetting agent to available fuel types in the compound, such as wood and paper. Since methylene chloride is a liquid at normal temperatures, it could have been absorbed into the furnishings of the compound. From my experiments, I can conclude that the methylene chloride had no enhancement effect on the fire spread over room furnishings. Also, I can conclude, from the flashpoint data (197°C or 387°F) of CS itself, that its deposition on furnishings should not have had a significant effect on fire propagation either. The tear gas had no bearing on the propagation of this fire.

Wind

Wind effects did have a profound effect on the external fire spread over the compound. An approximate 25-

mph wind from the south caused the fire plume to be bent approximately 65° from the vertical when the fire fully involved the compound. It is estimated that the fire was expending 3,600 MW at this time with an observed length of approximately 240 ft.

Wind effects did not appear to have had a significant effect on the fire growth within the compound, as seen in Fire 1 where flames and smoke emerge periodically from the right tower windows into the wind. This effect could have been as a result of closed doors or windows on the downwind side of the compound. The tank-made openings on the front of the compound could have had some effect on fire growth over the first floor, but more significantly could have provided air to areas of refuge for some of the occupants.

Survivability

It is estimated that the occupants would have had sufficient warning of the fire to enable them to escape, for at least up to 5 min from its inception, and up to nearly 20 min in some more protected locations. This is dramatically indicated by one occupant who jumps from the second floor 12 min after the start of the fire. Although smoke would have impaired visibility, exits were within 30 ft of most occupants, with additional openings made by the battering tanks.

Carbon monoxide in the smoke would have been the primary threat to the occupants. However, preliminary autopsy reports made available to me indicated that only five of thirty-one victims with recorded CO levels, had lethal levels of carbon monoxide (CO). The remaining twenty-six victims, with recorded CO data, stopped breathing before lethal CO levels were attained. If these data are correct, at least twenty-six victims did not die due to the fire. The autopsy report goes on to indicate that, in at least twenty-seven of the victims, the cause of death could be attributed to gunshot wounds.

Concluding Remarks

During the weeks preceding the fire at the Branch Davidian compound, we were all bystanders to the drama of the standoff and wondered how it would end. The eventual outcome was a horrible event. In the 2 years since, many theories about the fire have been proposed; some quite bizarre. I hope this presentation, our report, and the video I would like to submit, will help to explain the events of this fire.

The application of science to fire investigation can be very productive. Not only can it lead to information necessary in criminal and civil cases, but it can lead to changes in design. Such changes can be based on technical evidence, not judgment. Moreover, a fire laboratory in support of investigations can be very significant in acquiring and establishing data and evidence necessary to reconstruct the alleged events.

COMPUTER FIRE MODELS

Computer models are mathematical solutions implemented and displayed on computers. They attempt to alleviate the user from the direct development of the mathematical solution, and usually make it convenient and easy for the user. But in most all cases, the mathematics offers an approximate solution to a physical problem. The user may wish to solve a problem outside the scope of the original model. This attempt can lead to serious misinterpretations. Or the developer of the computer model may have made approximations or omissions not apparent to the user. The use of computer models requires an understanding of the problem and a knowledge of the model. It should only be a convenient tool, not a user-friendly device for any use.

zone model
a type of computer fire model that approximates the fire conditions in a room as two uniform layers with a source

We have already seen an example of the output of a computer model in Figures 10-2 and 10-3 (pages 202 and 203). These resulted from a fire model called a **zone model**, which is based on a representation of fire and smoke as shown in Figure 9-4 (page 175) where a hot upper layer (zone) is discriminated from a lower cooler zone. These two zones form the basis of a mathematical modeling concept for linking rooms in a building. The vent flows complete the linkage. This approach gives relatively coarse output for the building or room fire. A single temperature (or other property) is computed for each zone as the fire develops in time. In current models, the fire is prescribed. It is not predicted, but must be selected as an input, which places some limitations on the model's value. But with a good understanding of fire, the zone modeling approach can lead to impressive results. See Figures 10-2 and 10-3. As we understand how to model fire more completely, we can eliminate this limitation of selecting an input fire.

field model
a type of computer fire model that attempts to predict conditions at every point; also known as computational fluid dynamic models

Another form of fire model has been termed a **field model**, ultimately yielding predictions for temperature and other variables at each point and for each time. This model requires large and fast computers to conduct the many thousands of mathematical steps in its solution. These field models are generally called computational fluid dynamic (CFD) models. Usually they use the fundamental (exact) equations of physics and chemistry, but often they use special models to describe turbulence and combustion. They, too, usually prescribe the fire. However, the evolution of computers is allowing more direct solution of the exact equations. An example of such an output is shown in Figure 10-19, where it can be seen that the computation appears very realistic. It displays contours of constant temperatures at an instant in time that are colored in accordance with a real fire plume. The results are very impressive.

The student of fire computations must begin with an understanding of the processes associated with fire. Then the mathematical tools must be understood. Only then can the use of computer models truly benefit the student.

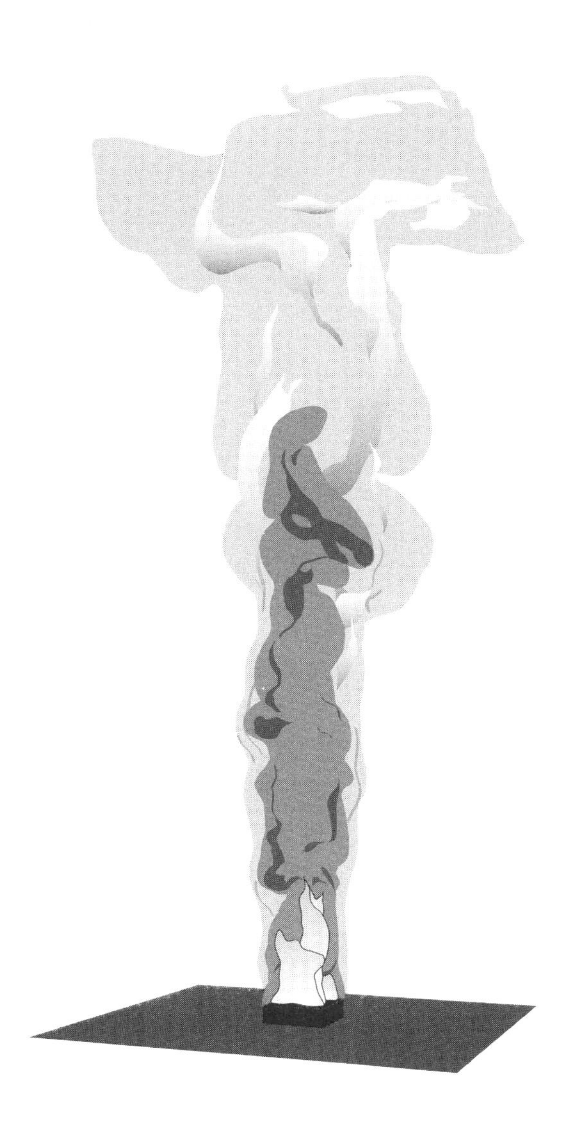

Figure 10-19 *Fire plume computed by a CFD model. From McGratten, Ref.7.*

Summary

A consideration of applications to fire safety design and fire investigation has been discussed. In both, a description of the fire growth in terms of $\dot{Q}(t)$ (kW) is needed to express the hazard and the damage. Fire safety is dealt with by regulations, and equivalency principles allow the introduction to engineering analysis. Many of the principles and processes quantitatively described in the previous chapters can be used in various design or investigative problems. Several examples of fire investigation were presented including the Branch Davidian fire near Waco, Texas, of April 19, 1993.

Computer models can be convenient useful tools in a fire analysis but do require an understanding of fire and its associated phenomena.

Review Question

1. Consider each of the examples presented in the chapter. Raise questions, comments, and other calculations to challenge or defend the conclusion.

Activities

1. Consider one or two examples from your fire experience to analyze. Have the class discuss these and work up calculations to quantify features of the problem. Discuss.
2. Identify computer models sold or made available to the fire community. Investigate their contents. Discuss.
3. Think of regulations in fire safety that could be enhanced by analyses. Discuss.
4. What additional aspects of fire safety would you like to see computed? Are these formulas available?

References

1. T. Tanaka, *A Model of Multiroom Fire Spread*, NBSIR 83-2718 (Gaithersburg, MD: National Bureau of Standards, August, 1983).
2. J. Yamaguchi, T. Fujita, T. Tanaka, and T. Wakamatsu, "A Study on Predicting Smoke Transport in a High-Rise Building" (in Japanese) in *Proceedings of the Annual Meeting* (Japanese Association for Fire Science and Engineering (JAFSE), May 1995), Japan.
3. T. J. Ohlemiller, "Smoldering Propagation on Solid Wood," in *Fire Safety Science, Proceedings of the Third International Symposium*, edited by G. Cox and B. Langford (London: Elsevier Applied Science, 1991), 565–574.
4. J. G. Quintiere, "Estimating Fire Growth on Compartment Interior Finish Materials," Honors Lecture, SFPE Engineering Seminars, Society of Fire

Protection Engineers, Spring Meeting, San Francisco, May 1994.

5. B. J. Hughes, private communications of statements and records, Office of the District Attorney, Kings County, Brooklyn, NY, 1994.
6. J. G. Quintiere, Statement on Matter of the Branch Davidians near Waco, Texas before the Subcommittee on Crime of the Committee on the Judiciary and the Subcommittee on National Security, International Affairs and Criminal Justice of the Committee on Governmental Reform and Oversight, House of Representatives, First Session, 104th Congress, July 26, 1995.
7. K. B. McGrattan, private communications, National Institute of Standards and Technology, Gaithersburg, MD, 1995.

Numbered Equations

CHAPTER 3 NUMBERED EQUATIONS

$$\dot{q} = kA(T_2 - T_1)/l \quad \textbf{(3-1)}$$

$$\dot{q}'' = h\,(T_2 - T_1) \quad \textbf{(3-2)}$$

$$\dot{q}'' = \sigma T^4 \quad \textbf{(3-3)}$$

$$\varepsilon = 1 - \exp\,(-\kappa l) \quad \textbf{(3-4)}$$

$$\dot{q}'' = \varepsilon\,\sigma\,T_2^4\,F_{12} \quad \textbf{(3-5)}$$

$$\dot{q}'' = \frac{X_r \dot{Q}}{4\pi c^2} \quad \textbf{(3-6)}$$

CHAPTER 4 NUMBERED EQUATIONS

$$T = T_\infty + \frac{\dot{q}'' t}{\rho c l} \quad \textbf{(4-1)}$$

$$t_{ig} = \rho c l \frac{(T_{ig} - T_\infty)}{\dot{q}''} \quad \textbf{(4-2)}$$

$$t_{ig} = C(k\rho c)\left[\frac{(T_{ig} - T_\infty)}{\dot{q}''}\right]^2 \quad (4\text{-}3)$$

CHAPTER 5 NUMBERED EQUATIONS

$$\rho VAc\,(T_{ig} - T_s) = \dot{q} \quad (5\text{-}1)$$

$$V = \frac{\dot{q}}{\rho cA(T_{ig} - T_s)} \quad (5\text{-}2)$$

$$\dot{q} = \dot{q}''\,\delta_f\,w \quad (5\text{-}3)$$

$$V = \frac{\dot{q}''\delta_f}{\rho cl(T_{ig} - T_s)} \quad (5\text{-}4)$$

$$V = \delta_f / t_{ig} \quad (5\text{-}5)$$

$$V = \frac{\phi}{k\rho c(T_{ig} - T_s)^2} \quad (5.6)$$

$$V = \frac{\dot{q}''}{\rho_b c(T_{ig} - T_s)} \quad (5\text{-}7)$$

$$V = C/\rho_b \quad (5\text{-}8)$$

$$V = (1 + V_\infty)C/\rho_b \quad (5\text{-}9)$$

CHAPTER 6 NUMBERED EQUATIONS

$$\dot{m}'' = \frac{\dot{q}''}{L} \quad (6\text{-}1)$$

$$\dot{Q} = \dot{m}'' A \Delta H_c \quad (6\text{-}2)$$

CHAPTER 7 NUMBERED EQUATIONS

$$f = 1.5/\sqrt{D} \quad (7\text{-}1)$$

$$\text{rate of air mass entrained} \sim \rho_a\,DL_f\,V_e \quad (7\text{-}2)$$

$$\dot{m}_f = \rho_f V_e \frac{\pi D^2}{4} \quad (7\text{-}3)$$

$$\frac{L_f}{D} \sim \frac{(\rho_f)s}{\rho_a} \quad (7\text{-}4)$$

$$V = \sqrt{\frac{2\,(T - T_a)gz}{T_a}} \quad (7\text{-}5)$$

$$Q^* = \frac{\dot{Q}}{\rho_a c_{pa} T_a \sqrt{gD}\,D^2} \quad (7\text{-}6)$$

$$Q^* = \frac{\dot{Q}\,(\mathrm{kW})}{1101\,[D(\mathrm{m})]^{5/2}} \quad (7\text{-}7)$$

$$Q^{*2/5} = \frac{(\dot{Q}/\rho_a c_{pa} T_a \sqrt{g})^{2/5}}{D} \quad (7\text{-}8)$$

$$L_f = 0.23\,\dot{Q}^{2/5} - 1.02\,D \quad (7\text{-}9)$$

$$\dot{Q} = \dot{m}\,c_p\,(T - T_a),\ (\mathrm{kW}) \quad (7\text{-}10)$$

$$\dot{Q} = \dot{m}_f\,\Delta H_c\,(1 - X_r),\ (\mathrm{kW}) \quad (7\text{-}11)$$

$$\dot{m} = n\,s\,\dot{m}_f \quad (7\text{-}12)$$

$$T - T_a = \frac{\Delta H_c(1 - X_r)}{c_p ns} \quad (7\text{-}13)$$

$$T - T_a = \frac{3000\,(1 - X_r)}{n} \quad \textbf{(7-14)}$$

$$\frac{T - T_a}{T_a} \sim \text{function}\ \frac{z}{(\dot{Q}/\rho_a c_{pa} T_a \sqrt{g})^{2/5}} \quad \textbf{(7-15)}$$

$$T - T_a = \frac{\dot{m}_f \Delta H_c (1 - X_r)}{\dot{m} c_p} \quad \textbf{(7-16)}$$

CHAPTER 8 NUMBERED EQUATIONS

$$\Phi = \left(\frac{\text{mass of fuel (gas)}}{\text{mass of air}}\right) \times s \quad \textbf{(8-1)}$$

$$y_{\text{co}} = \frac{m_{\text{co}}}{m} \quad \textbf{(8-2)}$$

$$\dot{m}_{\text{CO}} = y_{\text{CO}}\,\dot{m}''A, \quad \textbf{(8-3)}$$

$$s \approx \frac{\Delta H_{\text{c,ideal}}}{3\ \text{kJ/g}_{\text{air}}} \quad \textbf{(8-4)}$$

$$Y_{\text{species}} = \frac{\dot{m}_{\text{species}}}{\dot{m}_{\text{smoke}}} = \frac{Y_{\text{species}}\dot{m}}{\dot{m}_{\text{smoke}}} \quad \textbf{(8-5)}$$

$$X_{\text{species}} \approx \frac{29\,Y_{\text{species}}}{M_{\text{species}}} \quad \textbf{(8-6)}$$

$$\text{COHb}(\%) \approx 0.33\ RMV \cdot X_{\text{CO}}\ (\%) \cdot t(\text{min}) \quad \textbf{(8-7)}$$

$$I = I_O \exp(-\kappa_s l) \quad \textbf{(8-8)}$$

$$\kappa_s = \frac{\dot{m} D_m}{\dot{V}} \quad \textbf{(8-9a)}$$

$$\kappa_s = \frac{m D_m}{V} \quad \textbf{(8-9b)}$$

CHAPTER 9 NUMBERED EQUATIONS

$$\Delta p = (\rho_a - \rho) g H \quad \textbf{(9-1)}$$

$$V = \sqrt{\frac{2(\rho_a - \rho) g H}{\rho}} \quad \textbf{(9-2)}$$

$$\Delta p_{\text{fan}} = f \frac{H}{D} \frac{V^2}{2} \quad \textbf{(9-3)}$$

$$\Delta p_b = \rho_a (1 - T_a/T) g H \quad \textbf{(9-4)}$$

$$\dot{m}_{\text{a,max}} = 0.5 A_o \sqrt{H_o}\ (\text{kg/s}) \quad \textbf{(9-6)}$$

$$\Delta T\ (^\circ\text{C}) = 6.85 \left[\frac{\dot{Q}^2}{(hA)(A_o \sqrt{H_o})}\right]^{1/3} \quad \textbf{(9-7)}$$

$$\dot{Q}_{\text{FO}} = 624 \left[(hA)\left(A_o \sqrt{H_o}\right)\right]^{1/2} \quad \textbf{(9-8)}$$

$$\dot{Q}_{\max} = \dot{m}_{\text{a,max}}\ (\text{kg/s}) \times 3{,}000\ \text{kJ/kg} \quad \textbf{(9-9)}$$

Reference Tables

Type and Location	Number of Deaths	Date of Disaster
Floods:		
Galveston, Tx. tidal wave	6,000	Sept. 8, 1900
Johnstown, Pa.	2,209	May 31, 1889
Ohio and Indiana	732	Mar. 28, 1913
St. Francis, Calif., dam burst	450	Mar. 13, 1928
Ohio and Mississippi River valleys	380	Jan. 22, 1937
Hurricanes:		
Florida	1,833	Sept. 16–17, 1928
New England	657	Sept. 21, 1938
Louisiana	500	Sept. 29, 1915
Florida	409	Sept. 1–2, 1935
Louisiana and Texas	395	June 27–28, 1957
Tornadoes:		
Illinois	606	Mar. 18, 1925
Mississippi, Alabama, Georgia	402	Apr. 2–7, 1936
Southern and Midwestern states	307	Apr. 3, 1974
Ind., Ohio, Mich., Ill., and Wis.	272	Apr. 11, 1965
Ark., Tenn., Mo., Miss., and Ala.	229	Mar. 21–22, 1952
Earthquakes:		
San Francisco earthquake and fire	452	Apr. 18, 1906
Alaskan earthquake-tsunami hit Hawaii, Calif.	173	Apr. 1, 1946
Long Beach, Calif., earthquake	120	Mar. 10, 1933
Alaskan earthquake and tsunami	117	Mar. 27, 1964
San Fernando—Los Angeles, Calif., earthquake	64	Feb. 9, 1971
Marine:		
"Sultana" exploded—Mississippi River	1,547	Apr. 27, 1865
"General Slocum" burned—East River	1,030	June 15, 1904
"Empress of Ireland" ship collision—St. Lawrence River	1,024	May 29, 1914
"Eastland" capsized—Chicago River	812	July 24, 1915
"Morro-Castle" burned—off New Jersey coast	135	Sept. 8, 1934
Aircraft:		
Crash of scheduled plane near O'Hare Airport, Chicago	273	May 25, 1979
Aircraft (Continued):		
Crash of scheduled plane, Detroit, Mich.	156	Aug. 16, 1987
Crash of scheduled plane in Kenner, La.	154	July 9, 1982
Two-plane collision over San Diego, Calif.	144	Sept. 25, 1978
Crash of scheduled plane, Ft. Worth/Dallas Airport	135	Aug. 2, 1985
Railroad:		
Two-train collision near Nashville, Tenn.	101	July 9, 1918
Two-train collision, Eden, Colo.	96	Aug. 7, 1904
Avalanche hit two trains near Wellington, Wash.	96	Mar. 1, 1910
Bridge collapse under train, Ashtabula, Ohio	92	Dec. 29, 1876
Rapid transit train derailment, Brooklyn, N.Y.	92	Nov. 1, 1918
Fires:		
Peshtigo, Wis. and surrounding area, forest fire	1,152	Oct. 9, 1871
Iroquois Theatre, Chicago	603	Dec. 30, 1903
Northeastern Minnesota forest fire	559	Oct. 12, 1918
Cocoanut Grove nightclub, Boston	492	Nov. 28, 1942
North German Lloyd Steamships, Hoboken, N.J.	326	June 30, 1900
Explosions:		
Texas City, Texas, ship explosion	552	Apr. 16, 1947
Port Chicago, Calif., ship explosion	322	July 18, 1944
New London, Texas, school explosion	294	Mar. 18, 1937
Oakdale, Pa., munitions plant explosion	158	May 18, 1918
Eddystone, Pa., munitions plant explosion	133	Apr. 10, 1917
Mines:		
Monongha, W. Va., coal mine explosion	361	Dec. 6, 1907
Dawson, N. Mex., coal mine fire	263	Oct. 22, 1913
Cherry, Ill., coal mine fire	259	Nov. 13, 1909
Jacobs Creek, Pa., coal mine explosion	239	Dec. 19, 1907
Scofield, Utah, coal mine explosion	200	May 1, 1900

Source: World Almanac, National Transportation Safety Board, National Weather Service, National Fire Protection Association, Chicago Historical Society, American Red Cross, U.S. Bureau of Mines, National Oceanic and Atmospheric Administration, and city and state Boards of Health.

Figure 1-2 *Life loss due to national disasters in the United States. Courtesy National Safety Council,* Accident Facts, *1995 Edition (Ref. 2).*

Table 1-1 *Annual fire death rates.*

Country	Annual Deaths per 10^5 persons	Country	Annual Deaths per 10^5 persons
Russia*	10.60	France	1.26
Hungary	3.31	Czech Republic	1.21
India*	2.20	Germany	1.17
Finland	2.18	Australia	0.93
Union of South Africa*	2.00	New Zealand	0.92
United States	1.95	Spain	0.86
Denmark	1.64	Poland*	0.80
Norway	1.60	Austria	0.74
Canada	1.58	Netherlands	0.63
Japan	1.52	Switzerland	0.53
United Kingdom	1.49	Italy*	0.30
Belgium	1.47	China*	0.20
Sweden	1.35		

Sources: From Wilmot, Ref. 4, for 1989–1992. Starred items are from Brushlinsky et al., Ref. 5 for 1994.

Table 1-2 *National annual fire costs by percentage of gross domestic product for 1989 to 1991.*

Country	Property Loss	Building Fire Protection	Fire Insurance Admin.	Fire Fighting	Total
Hungary	0.12	0.42	0.01	ND	ND
Spain	0.12	ND	0.05	ND	ND
Japan	0.14	0.29	0.11	0.27	0.81
Finland	0.17	ND	0.05	0.18	ND
U.S.A.	0.17	0.37	0.07	0.28	0.89
Canada	0.18	0.18	0.21	0.16	0.73
New Zealand	0.18	0.14	0.23	0.20	0.75
West Germany	0.19	ND	0.09	ND	ND
Netherlands	0.20	0.12	0.04	0.17	0.53
Austria	0.21	ND	0.14	ND	ND
United Kingdom	0.22	0.19	0.10	0.27	0.78
Switzerland	0.23	0.29	ND	ND	ND
Denmark	0.28	ND	0.08	0.09	ND
Sweden	0.28	0.10	0.06	0.19	0.63
France	0.29	0.14	0.16	ND	ND
Norway	0.31	0.26	0.15	0.12	0.84
Belgium	0.40	0.21	0.28	0.18	1.07

Source: From Wilmot Ref. 4.

ND = data not available.

Table 1-3 *SI quantities.*

Quantity	Unit abbreviation
Force	N (newton)
Mass	kg (kilogram mass)
Time	s (second)
Length	m (meter)
Temperature	°C or K
Energy	J (joule)
Power	W (watt)
Thermal conductivity	W/m - °C
Heat-transfer coefficient	W/m^2 - °C
Specific heat	J/kg - °C
Heat flux	W/m^2

Table 1-4 *Alternative energy units.*

1 Btu will raise 1 lb_m of water 1°F at 68°F.

1 cal will raise 1 g of water 1°C at 20°C.

1 kcal will raise 1 kg of water 1°C at 20°C.

Some conversion factors for the various units of work and energy are

1 Btu = 778.16 lb_f-ft

1 Btu = 1055 J

1 kcal = 4182 J

1 lb_f-ft = 1.356 J

1 Btu = 252 cal

Table 1-5 *Temperature conversions.*

°F degree Fahrenheit: T(F) = T(C) (1.8) + 32

°R degree Rankine: T(R) = T(F) + 459.69

°C degree Celsius or Centigrade: T(C) = (T(F)–32)/1.8

°K degree Kelvin: T(K) = T(C) + 273.16

Table 1-6 *Common conversion factors and symbols.*

length	1 m = 3.2808 ft	l
area	1 m^2 = 10.7639 ft^2	A
density	1 kg/m^3 = 0.06243 lb/ft^3	ρ
energy	1 kJ = 0.94783 Btu	Q
heat	1 kJ = 0.94783 Btu	q
heat flow rate	1 W = 3.4121 Btu/hr	$\dot{q}$
energy release rate	1 W = 3.4121 Btu/hr	$\dot{Q}$
heat flow rate per unit area, heat flux	1 W/cm^2 = 0.317 Btu/hr-ft^2 1 W/cm^2 = 10.kW/m^2	$\dot{q}''$
specific heat	1 kJ/kg-°C = 0.23884 Btu.lb-°F	c
thermal conductivity	1 W/m-°C = 0.5778 Btu/hr-ft-°F	k
thermal diffusivity	1 m^2/s = 10.7639 ft^2/s	α
pressure	1 atm = 14.69595 lb_f/in^2 = 1.01325 × 10^5 N/m^2 1 N/m^2 = 1 Pascal (Pa)	P

Table 1-7 *Scientific notations, prefixes.*

Multiplier	Prefix	Abbreviation
10^{12}	tera	T
10^{9}	giga	G
10^{6}	mega	M
10^{3}	kilo	k
10^{2}	hecto	h
10^{-2}	centi	c
10^{-3}	milli	m
10^{-6}	micro	μ
10^{-9}	nano	n
10^{-12}	pico	p
10^{-18}	atto	a

Note: For example, 10^3 = 1000 and 10^{-3} = 0.001.

Table 2-1 *Typical smolder velocities associated with fuel configuration.*

Fuel	Fuel/Smolder Configuration	Air Supply Condition/Rate	Smolder Velocity (cm/sec)
Pressed fiber insulation board, 0.23–0.29 g/cc	1.3 cm thick, horizontal strips, width large compared to thickness	Natural convection/diffusion	$1.3–2.2 \times 10^{-3}$
Pressed fiber insulation board, 0.23–0.29 g/cc	1.3 cm × 1.3 cm strips varied angle to vertical	Natural convection/diffusion	$2.7–4.7 \times 10^{-3}$
Pressed fiber insulation board, 0.23–0.29 g/cc	1.3 cm × 5 cm strips forward smolder	Forced flow, 20 to 1500 cm/s	3.5×10^{-3} (in 20 cm/s air) 13.0×10^{-3} (in 1400 cm/s air)
Pressed fiber insulation board, 0.23–0.29 g/cc	1.3 cm × 5 cm strips reverse smolder	Forced flow, 80–700 cm/s	$2.8–3.5 \times 10^{-3}$
Pressed fiberboard (pine or aspen) 0.24 g/cc	1.3 cm × 30 cm sheets, horizontal, forward smolder	Forced flow, 10–18 cm/s	0.7×10^{-3}
Cardboard	Vertical rolled cardboard cylinder, downward propagation, varied dia. 0.19–0.38 cm	Natural convection, diffusion	$5.0–8.4 \times 10^{-3}$
Shredded tobacco	0.8 cm dia. cigarette, horizontal, in open air	Natural convection, diffusion	$3.0–5.0 \times 10^{-3}$
Cellulose fabric + 3% NaCl	Double fabric layer, 0.2 cm thick, horizontal, forward smolder	Forced flow, ≈ 10 cm/s	$\approx 1.0 \times 10^{-2}$

Source: After Ohlemiller, Ref. 4.

Table 2-2 *Sawdust cubes, with and without oil, exposed to air temperatures.*

Cube Size 2r mm	Oil Content (%)	Ignition Temperature (°C)
25.4	0	212
25.4	11.1	208
51	0	185
51	11.1	167
76	0	173
76	11.1	146
152	0	152
152	11.1	116
303	0	135
303	11.1	99
910	0	109
910	11.1	65

Source: From Bowes, Ref. 5.

Table 2-3 *Flammability limits of gaseous fuels in air at normal atmospheric temperature and pressure.*

	LFL (%)	UFL (%)	AIT (°C)
Acetylene	2.5	100	305
Benzene	1.3	7.9	560
n-Butane	1.8	8.4	405
Carbon Monoxide	12.5	74	609
Ethylene	2.7	36	490
n-Heptane	1.05	6.7	215
Hydrogen	4.0	75	400
Methane	5.0	15.0	540
Propane	2.1	9.5	450
Trichlorethylene	12	40	420

Source: Based on data from Beyler, Ref. 7.

Table 2-4 *Detonation limits at normal atmospheric temperature and pressure.*

	Lower Limit (%)	Upper Limit (%)	Detonation Velocity (m/s)
Hydrogen in pure O_2	15	90	2821
Hydrogen in air	18.3	59	—
CO in pure moist O_2	38	90	1264
Propane in pure O_2	3.2	37	2280, 2600
Acetylene in air	4.2	50	—
Acetylene in pure O_2	3.5	92	2716

Source: From Lewis and von Elbe, Ref. 8.

Table 3-1 *Table of thermal properties.*

Material	Thermal Conductivity (k) (W/m-K)	Specific Heat (c) (kJ/kg-K)	Density (ρ) (kg/m^3)	Thermal Diffusivity (α) (m^2/s)	Thermal Inertia ($k\rho c$) (kW^2-s/m^4-K^2)
Copper	387	0.380	8940	1.14×10^{-4}	1300.0
Steel (mild)	45.8	0.460	7850	1.26×10^{-5}	160.0
Brick (common)	0.69	0.840	1600	5.2×10^{-7}	0.93
Concrete	0.8–1.4	0.880	1900–2300	5.7×10^{-7}	2.0
Glass (plate)	0.76	0.840	2700	3.3×10^{-7}	1.7
Gypsum plaster	0.48	0.840	1440	4.1×10^{-7}	0.58
PMMA	0.19	1.420	1190	1.1×10^{-7}	0.32
Oak	0.17	2.380	800	8.9×10^{-8}	0.32
Yellow pine	0.14	2.850	640	8.3×10^{-8}	0.25
Asbestos	0.15	1.050	577	2.5×10^{-7}	0.091
Fiber insulating board	0.041	2.090	229	8.6×10^{-8}	0.020
Polyurethane foam	0.034	1.400	20	1.2×10^{-6}	9.5×10^{-4}
Air	0.026	1.040	1.1	2.2×10^{-5}	3.0×10^{-5}

Source: From Drysdale, Ref. 1.

Table 3-2 *Typical values for convective coefficients,* h

Fluid Condition	h (W/m^2°C)
Buoyant flows in air	5–10
Laminar match flame	~ 30
Turbulent liquid pool fire surface	~ 20
Fire plume impinging on a ceiling	5–50
2 m/s wind speed in air	~ 10
35 m/s wind speed in air	~ 75

Table 3.3 *Typical radiative energy fractions, X_r.*

Fuel (l > 0.5 m)	X_r (%)
Methanol, methane	15–20
Butane, benzene, wood cribs	20–40
Hexane, gasoline, polystyrene	40–60

Source: Various.

Table 3-4 *Specific radiation fraction of combustion energy for hydrocarbon pool fires showing dependence on diameter.*

Hydrocarbon	Pool Size (m)	% Radiative Output/Combustion Output
Methanol	1.2	17.0
LNG on land	18.0	16.4
	0.4 to 3.05	15.0 to 34.0
	1.8 to 6.1	20.0 to 25.0
	20.0	36.0
LNG on water	8.5 to 15.0	12.0 to 31.0[a]
LPG on land	20.0	7.0
Butane	0.3 to 0.76	19.9 to 26.9
Gasoline	1.22 to 3.05	40.0 to 13.0[a]
	1.0 to 10.0	60.1 to 10.0[a]
Benzene	1.22	36.0 to 38.0
Hexane	—	40
Ethylene	—	38

[a]In these cases, the smaller diameter fires were associated with higher radiative outputs.

Source: From Mudan and Croce, Ref. 3.

Table 4-1 *Critical temperatures for liquid fuels.*

Liquid	Formula	T_{FP} (K)	T_B (K)	T_{AUTO} (K)[a]
Propane	C_3H_8	169	231	723
Gasoline	mixture	~ 228	~ 306	~ 644
Acrolein	C_3H_4O	247	326	508
Acetone	C_3H_6O	255	329	738
Methanol	C_3H_3OH	285	337	658
Ethanol	C_2H_5OH	286	351	636
Kerosene	~ $C_{14}H_{30}$	~ 322	~ 505	~ 533
m-Creosol	C_7H_8O	359	476	832
Formaldehyde	$C H_2O$	366	370	703

[a]Based on a stoichiometric mixture in a vessel.

Table 4-2 *Typical ignition times of thick solids.*

Heat Flux (kW/m²)	Time (s)	Material
10	300	Plexiglas, polyurethane foam, acrylate carpet
20	70	Wool carpet
	150	Paper on gypsum board
	250	Wood particleboard
30	5	Polyisocyanurate foam
	70	Wool/nylon carpet
	150	hardboard

Table 4-3 *Ignition properties.*

Material	$k\rho c$ $(kW/m^2K)^2s$	T_{ig} (°C)	$\dot{q}''_{critical}$ (kW/m^2)
Plywood, plain (0.635cm)	0.46	390.	16.
Plywood, plain (1.27cm)	0.54	390.	16.
Plywood, FR (1.27cm)	0.76	620.	44.
Hardboard (6.35mm)	1.87	298.	10.
Hardboard (3.175mm)	0.88	365.	14.
Hardboard (gloss paint), (3.4mm)	1.22	400.	17.
Hardboard (nitrocellulose paint)	0.79	400.	17.
Particleboard (1.27cm stock)	0.93	412.	18.
Douglas Fir particleboard (1.27cm)	0.94	382.	16.
Fiber insulation board	0.46	355.	14.
Polyisocyanurate (5.08cm)	0.020	445.	21.
Foam, rigid (2.54cm)	0.030	435.	20.
Foam, flexible (2.54cm)	0.32	390.	16.
Polystyrene (5.08cm)	0.38	630.	46.
Polycarbonate (1.52mm)	1.16	528.	30.
PMMA type G (1.27cm)	1.02	378.	15.
PMMA polycast (1.59mm)	0.73	278.	9.
Carpet #1 (wool, stock)	0.11	465.	23.
Carpet #2 (wool, untreated)	0.25	435.	20.
Carpet #2 (wool, treated)	0.24	455.	22.
Carpet (nylon/wool blend)	0.68	412.	18.
Carpet (acrylic)	0.42	300.	10.
Gypsum board, (common) (1.27mm)	0.45	565.	35.
Gypsum board, FR (1.27cm)	0.40	510.	28.
Gypsum board, wall paper	0.57	412.	18.
Asphalt shingle	0.70	378.	15.
Fiberglass shingle	0.50	445.	21.
Glass reinforced polyester (2.24mm)	0.32	390.	16.
Glass reinforced polyester (1.14mm)	0.72	400.	17.
Aircraft panel, epoxy fiberite	0.24	505.	28.

Source: From Quintiere and Harkleroad, Ref. 4.

Table 5-1 *Lateral flame spread data from ASTM E 1321.*

Material	T_{ig} (°C)	$k\rho c$ $\left(\frac{kW}{m^2K}\right)^2 s$	ϕ $\left(\frac{kW^2}{m^3}\right)$	$T_{s,min}$ (°C)
Wood fiber board	355	0.46	2.3	210.
Wood hardboard	365	0.88	11.0	40.
Plywood	390	0.54	13.0	120.
PMMA	380	1.0	14.4	<90.
Flexible foam plastic	390	0.32	11.7	120.
Rigid foam plastic	435	0.03	4.1	215.
Acrylic carpet	300	0.42	9.9	165.
Wallpaper on plasterboard	412	0.57	0.8	240.
Asphalt shingle	378	0.70	5.4	140
Glass reinforced plastic	390	0.32	10.0	80

Source: From Quintiere and Harkleroad, Ref. 2.

Table 5.2 *Typical flame spread rates.*

Spread	Rate (cm/s)
Smoldering	0.001 to 0.01
Lateral or downward spread on thick solids	0.1
Wind driven spread through forest debris or brush	1 to 30
Upward spread on thick solids	1.0 to 100.
Horizontal spread on liquids	1.0 to 100.
Premixed flames	10. to 100. (laminar) $\approx 10^5$ (detonations)

Table 5.3 *Empirical standard flame spread tests.*

ASTM Designation: E 84	Standard Test Method for Surface Burning Characteristics of Building Materials
ASTM Designation: E 162	Standard Test Method for Surface Flammability of Materials Using a Radiant Heat Energy Source
ASTM Designation: E 648	Standard Test Method for Critical Radiant Flux of Floor-Covering Systems Using a Radiant Heat Energy Source
ASTM Designation: E 1321	Standard Test Method for Determining Material Ignition and Flame Spread Properties

Table 6-1 *Heat of gasification values.*

Fuel	L (kJ/g)
Liquids:	
Gasoline	0.33
Hexane	0.45
Heptane	0.50
Kerosene	0.67
Ethanol	1.00
Methanol	1.23
Thermoplastics	
Polyethylene	1.8–3.6
Polypropylene	2.0–3.1
Polymethylmethacrylate	1.6–2.8
Nylon 6/6	2.4–3.8
Polystyrene foam	1.3–1.9
Flexible polyurethane foam	1.2–2.7
Char Formers	
Polyvinyl chloride	1.7–2.5
Rigid polyurethane foam	1.2–5.3
Whatman filter paper no.3	3.6
Corrugated paper	2.2
Woods	4–6.5

Sources: Data from Tewarson and Quintiere et al., Refs. 1 and 2.

Table 6-2 *Maximum burning flux values.*

Fuel	$\dot{m}''$ (g/m²-s)
Liquified propane	100–130
Liquified natural gas	80–100
Benzene	90
Butane	80
Hexane	70–80
Xylene	70
JP-4	50–70
Heptane	65–75
Gasoline	50–60
Acetone	40
Methanol	22
Polystyrene (granular)	38
Polymethyl methacrylate (granular)	28
Polyethylene (granular)	26
Polypropylene (granular)	24
Rigid polyurethane foam	22–25
Flexible polyurethane foam	21–27
Polyvinyl chloride (granular)	16
Corrugated paper cartons	14
Wood crib	11

Source: From Tewarson, Ref. 1.

Table 6-3 *Effective heat of combustion, ΔH_c (kJ/g).*

Methane	50.0
Ethane	47.5
Ethene	50.4
Propane	46.5
Carbon monoxide	10.1
n-Butane	45.7
c-Hexane	43.8
Heptane	44.6
Gasoline	43.7
Kerosene	43.2
Benzene	40.0
Acetone	30.8
Ethanol	26.8
Methanol	19.8
Polyethylene	43.3
Polypropylene	43.3
Polystyrene	39.8
Polycarbonate	29.7
Nylon 6/6	29.6
Polymethyl methacrylate	24.9
Polyvinyl chloride	16.4
Cellulose	16.1
Glucose	15.4
Wood	13–15

Source: Based on data from Tewarson, Ref. 1.

Table 6-4 *Typical peak burning rate values (in g/s).*

Small waste containers (18–40ℓ)	3–6
Large waste containers (70–120ℓ)	5–10
Chairs, wood and upholstered	10–60
Sofas	20–100
Beds	20–140
Closet	~40
Office	~90
Bedroom	~130
Kitchen	~190
House	~30,000

Source: Based on data from Ref. 6.

Table 6-5 *Fire behavior of warehouse commodities: fully involved energy release rates for a fixed floor area.*

Commodity	$\dot{Q}$/Floor Area Covered (MW/m²)
Methanol	0.72
Diesel oil	1.9
Kerosene	2.2
Gasoline	2.2
Wood pallets, stacked 1 1/2 ft high	1.3
Wood pallets, stacked 5 ft high	3.7
Wood pallets, stacked 10 ft high	6.6
Wood pallets, stacked 15 ft high	9.9
Mail bags, filled, stored 5 ft high	0.39
PE letter trays, filled, stacked 5 ft high	8.2
PS insulation board, rigid foam, stacked 14 ft high	3.1
PU insulation board, rigid foam, stacked 15 ft high	1.9
PS tubs rested in cartons, stacked 14 ft high	5.1
FRP shower stalls in cartons, stacked 15 ft high	1.2
PE bottles in cartons, stacked 15 ft high	1.9
PS toy parts in cartons, stacked 15 ft high	2.0
PE trash barrels in cartons, stacked 15 ft high	2.9
Cartons, compartmented, stacked 15 ft high	2.2
PVC bottles packed in cartons, compartmented, stacked 15 ft high	3.3
PP tubs packed in cartons, compartmented, stacked 15 ft high	4.2
PE bottles packed in cartons, compartmented, stacked 15 ft high	6.1
PS jars packed in cartons, compartmented, stacked 15 ft high	14.0

Source: Based on data from Heskestad, Ref. 7.

PE = polyethylene, PU = polyurethane, PVC = polyvinyl chloride, PS = polystyrene, PP = polypropylene, FRP = fiberglass-reinforced-polyester

Table 6-6 *Characteristic times to reach 1 MW for t^2 fires.*

Commodity	t_1 (s)
Wood pallets, stacked 1 1/2 ft high	155–310
Wood pallets, stacked 5 ft high	92–187
Wood pallets, stacked 10 ft high	77–115
Wood pallets, stacked 16 ft high	72–115
Mail bags, filled, stored 5 ft high	187
Cartons, compartmented, stacked 15 ft high	58
Paper, vertical rolls, stacked 20 ft high	16–26
Cotton, polyester garments in 12 ft high rack	21–42
"Ordinary combustibles" rack storage, 15–30 ft high	39–262
Paper products, densely packed in cartons, rack storage, 20 ft high	461
PE letter trays, filled, stacked 5 ft high on cart	189
PE trash barrels in cartons, stacked 15 ft high	53
PE bottles packed in compartmented cartons, 15 ft high	82
PE bottles in cartons, stacked 15 ft high	72
PE pallets, stacked 3 ft high	145
PE pallets, stacked 6–8 ft high	31–55
PU mattress, single, horizontal	115
PU insulation board, rigid foam, stacked 15 ft high	7
PS jars packed in compartmented cartons, 15 ft high	53
PS tubs nested in cartons, stacked 15 ft high	115
PS insulation board, rigid foam, stacked 14 ft high	6
PUS bottles packed in compartmented cartons, 15 ft high	8
PP tubs packed in compartmented cartons, 15 ft high	9
PP and PE film in rolls, stacked 14 ft high	38
Distilled spirits in barrels, stacked 20 ft high	24–39

Source: Based on data from Heskestad, Ref. 7.

PE = polyethylene, PU = polyurethane, PVC = polyvinyl chloride, PS = polystyrene, PP = polypropylene, FRP = fiberglass-reinforced-polyester

Table 8-1 *Fuel properties as a function of ventilation.*

Fuel	Overventilated Conditions						Underventilated Conditions		
	Y_{CO_2} (g/g)	Y_{CO} (g/g)	Y_{soot} (g/g)	ΔH_c (kJ/g)	$\Delta H_{c,\ ideal}$ (kJ/g)	D_m (m²/g)	Y_{CO} (g/g)	Y_{H_2} (g/g)	Y_{HCl} (g/g)
Gases:									
Propane	2.85	0.005	0.024	43.7	46.4	0.155	0.229	0.011	—
Acetylene	2.6	0.042	0.096	36.7	48.2	0.315	NA*	NA	—
Liquids:									
Ethyl alcohol	1.77	0.001	0.008	25.6	26.8	NA	0.219	0.0098	—
Heptane	2.85	0.010	0.037	41.2	44.6	0.190	NA	NA	—
Solids:									
Wood (Red oak, pine)	1.27	0.004	0.015	12.4	17.7	0.037	0.138	0.0024	—
Polymethyl methacrylate (PMMA)	2.12	0.010	0.022	24.2	25.2	0.109	0.189	0.0032	—
Polystyrene (PS)	2.33	0.060	0.164	27.0	39.2	0.335	NA	NA	—
Nylon	2.06	0.038	0.075	27.1	30.8	0.230	NA	NA	—
Polyurethane (PU) -flexible foam	1.51	0.031	0.227	19.0	27.2	0.326	NA	NA	—
Polyvinyl chloride (PVC)	0.46	0.063	0.172	5.7	16.4	0.400	0.36	NA	0.400

Source: Based on data from Tewarson, Ref. 1.

*NA = not available

Table 8-2 *Effects of O_2 loss in blood.*

Due to Decrease in Oxygen Concentration		Due to Increase in Carbon Monoxide Concentration		
O_2Hb (%)	X_{O_2} (%)	COHb (%)	X_{CO} (%)	Effect
90–100	15–21	0–10	<0.008	None
80–90	12–15	10–20	0.008–0.015	Fatigue
60–80	8–12	20–40	0.015–0.04	Dizziness, nausea, possible paralysis
50–60	6–8	40–50	0.04–0.06	Prostration, asphyxiation, collapse
30–50	3–6	50–70	0.06–0.3	Unconscious in minutes, possible death
0–30	0–3	70–100	>0.3	Unconscious in seconds, death likely

Table 8-3 *Respiration minute volume (RMV) (liters/min.) for a 70-kg man.*

RMV	Activity
8.5	Resting
25	Light work
50	Heavy work, slow running

Source: Based on data from Purser, Ref. 2.

Table 8-4 *Effects of CO_2 on humans at normal atmospheric pressure and O_2 concentration.*

CO_2 Concentration in Inhaled Air (%)	Effect
0.04	Normal air
0.5	Safe limit, prolonged exposure
1.8–2.0	30–50% increase in ventilation rate
2.5–3.0	100% increase in ventilation rate
4.0	300% increase in ventilation rate
5.0	Dizziness, poisoning symptoms, > 30 minutes
7.0–9.0	Unconscious, in 15 minutes
10.0–30.0	Unconscious, in < 10 minutes, followed by death

Source: From Purser, Ref. 2.

Table 8-5 *Tolerance times under heat stress conditions.*

Exposure Temperature °C	RH %	Tolerance Time
49	10	~10 days
49	50	~2 hours
49	100	~10 minutes
100	0–100	~10 minutes

Source: From Purser, Ref. 2.

Table 9-1 *Typical construction properties.*

Insulation board	$k\rho c$ (kW^2 - s/m^4 - °C^2) 0.09	k (kW/m - C) 4.1×10^{-5}
Wood	0.30	1.5×10^{-4}
Gypsum board	0.60	5×10^{-4}
Concrete	2.0	1×10^{-3}
Steel	150	5×10^{-2}

Acronyms

AIT	autoignition temperature
ASTM	American Society of Testing and Materials
BATF	Bureau of Alcohol, Tobacco, and Firearms
BTU	British thermal unit
CDT	Central Daylight Time
CFD	computational fluid dynamics
FLETC	Federal Law Enforcement Training Center
FLIR	forward-looking infrared
GDP	gross domestic product
LFL	lower flammable limit
NBS	National Bureau of Standards
NFPA	National Fire Protection Association
NIST	National Institute for Standards and Technology
PMMA	polymethyl methacrylate
PRC	Product Research Committee
PS	polystyrene
PU	polyurethane
PVC	polyvinyl chloride
RH	relative humidity
RMV	respiration minute volume
SI	International System of Units
UFL	upper flammable limit

Glossary

Absorption coefficient That property that pertains to the amount of radiation absorbed per unit length.

Adiabatic flame temperature The maximum possible temperature in the reacting zone with no heat lost.

ASTM E 84 A standard providing a test (Steiner tunnel test) developed by Underwriters Laboratories to examine wind-aided flame spread under ceiling mounted materials.

ASTM E 162 A standard providing a radiant fuel test to measure the flammability of materials under downward spread.

ASTM E 648 A standard providing a flame test for floor-covering materials.

ASTM E 1321 A standard providing a test to determine data for ignition and flame spread.

Autoignition Initiation of fire by chemical process inherent in the material; specific fuel concentration and temperature is usually needed.

Autoignition temperature The lowest temperature at which a mixture of fuel and oxidizer can propagate a flame without the aid of an initiating energy source (pilot).

Backdraft The sudden eruption of fire in a compartment due to the accumulation of fuel gases and the introduction of fresh air.

Black body A radiator emitting the maximum possible energy.

Boiling point The maximum temperature at which a liquid can evaporate under normal atmospheric conditions; equilibrium temperature for a liquid and its vapor to coexist at 1 atmosphere of pressure.

Buoyancy An effective force on fluid due to density or temperature differences in a gravitational field.

Burning rate, $\dot{m}$ The mass of fuel consumed in the fire per unit time.

Burnout Point at which flames cease.

Carboxyhemoglobin Carbon-monoxide-bearing hemoglobin.

Characteristic combustion length $\left(\frac{\dot{Q}}{\rho_a c_p T_a \sqrt{g}}\right)^{2/5}$ A length scale representative of the fire size.

Charring The production of a solid carbonaceous residue on heating or burning a solid.

Chemical equation An equation showing the proportions (by mass) of the fuel, oxygen, and products in a chemical reaction.

Chemical kinetics Refers to the rate of the chemical reaction.

Combustion Fire, or controlled fire.

Completeness Pertaining to a combustion process going to its most stable state, an ideal reaction; water and carbon dioxide would be the complete products of combustion for hydrocarbon fires.

Concentration The percentage of material per unit mass (or volume) of its mixture.

Conduction Heat transfer due to molecular energy transfer following Fourier's Law.

Configuration factor Fraction of radiation received by a target compared to the total emitted by the source.

Conflagration or mass fire A fire over a large tract of land where generally the flames are much shorter than the horizontal extent of the fire.

Convection Conduction heat transfer from a moving fluid (gas or liquid) to a solid surface.

Convective heat transfer coefficient A quantity that represents the ability of heat to be transformed from a moving fluid to a solid surface expressed in terms of heat flux per unit temperature difference.

Cracking Pyrolysis; breaking gaseous molecules into other molecules.

Critical heat flux A threshold level of heating below which ignition (or in another context, flame spread) is not possible.

Detonation A premixed flame preceded by a shock wave.

Developing fire The early stage of growth (in a compartment fire) before flashover and full involvement.

Diffusion Process of species transport in a mixture from its high to low concentration.

Diffusion flame A flame in which the fuel and oxygen are transported (diffused) from opposite sides of the reaction zone (flame).

Dimensionless Having no units of measure (terms combine to produce no units).

Dose The accumulation of product concentration over time; the integral of concentration over time.

Eddies Rotating regions of a fluid.

Emissivity That property (0 to 1) that gives the fraction of being a perfect radiator.

Energy A state of matter representative of its ability to do work or transfer heat.

Energy release rate, $\dot{Q}$ The energy produced by the fire per unit time; fire power.

Entrainment The process of air or gases being drawn into a fire, plume, or jet.

Equivalence ratio, Φ The ratio of fuel to air times the stoichiometric air to fuel ratio, *s*; or $(\text{fuel/air})_{\text{available}}$ divided by $(\text{fuel/air})_{\text{stoichiometric}}$.

Equivalency The statement in regulations allowing for alternatives by design.

Evaporation The process of gas molecules escaping from the surface of a liquid.

Field model A type of computer fire model that attempts to predict conditions at every point; also known as a computational fluid dynamics model.

Fire An uncontrolled chemical reaction producing light and sufficient energy.

Fire plume The flame and gases emanating from a burning object.

Fire spread The process of an advancing fire front: smoldering or flaming.

Fire triangle A concept describing fire as consisting of three ingredients: fuel, oxygen, and energy.

Flame height The vertical measure of the combustion region.

Flame spread The process of advancing the fire front in air, along surfaces, or through porous solids.

Flashover A dramatic event in a room fire that rapidly leads to full involvement; an event that can occur at a smoke temperature of 500 to 600°C.

Flashpoint The temperature of a liquid fuel, theoretically corresponding to the lower flammable limit of its evaporated vapor, and the point of piloted ignition.

Fluid mechanics The study of fluid motion.

Flux Pertains to mass or heat flow rates per unit area.

Forced flow Refers to air flow produced by wind or a fan.

Free-burning Burning in open-air.

Frequency Cycles per unit time (measured in hertz, cycles per second).

Fuel lean Description of fuel burning in an excess supply of air.

Fuel-limited State of a compartment fire where the air supply is sufficient to maintain combustion.

Fully developed State of a compartment fire during which the flames fill the room involving all the combustibles, or the state of maximum possible energy release in a room fire.

Heat Energy transfer due to temperature difference.

Heat flux Heat flow rate per unit area of flow path.

Heat of combustion The energy released by the fire per unit mass of fuel burned.

Heat of gasification, *L* Energy required to produce a unit mass of fuel vapor from a solid or liquid.

Heat transfer The transport of energy from a high- to a low-temperature object.

Hemoglobin Compound in blood that transports oxygen or carbon monoxide.

Humidity The property of the water-air mixture that measures the amount of water present relative to the equilibrium concentration.

Hyperthermia Heat stress.

Ignition temperature The surface temperature needed to cause ignition in solids.

Incomplete combustion A combustion process that does not go to the most stable species such as H_2O and CO_2.

International System of Units or Standard International (SI) Units The system of units for measurement adopted by the science community.

Irritant gases Acid gases and other hydrocarbon byproducts that can cause pain on contact or inhalation.

Jet flame Flame due to a high-velocity fuel supply.

Kelvin (K) Absolute Celsius temperature scale, 273 + °C.

Laminar Refers to orderly, unfluctuating fluid motion.

Line fires Elongated fires on a horizontal fuel surface.

Mass burning flux, $\dot{m}''$ Burning rate per unit area.

Mass concentration, Y Ratio of species mass to mixture mass; also referred to as mass fraction.

Mass loss rate The mass of fuel vaporized but not necessarily burned per unit time.

Mass optical density, D_m Optical property related to the yield of particulates in smoke.

Narcosis Effect of inducing sleep.

Natural flow Refers to air flow induced by buoyancy.

Neutral plane The height above which smoke will or can flow out of a compartment; the height of zero pressure difference across a partition.

Newtons Second Law of Motion Relates force on a body to its mass and resulting acceleration.

Opposed-flow Refers to air flow in the direction opposite to the fire spread.

Overventilated More than stoichiometric air is available.

Oxygen bomb A device for measuring the maximum energy released in combustion for a given mass of fuel.

Oxygen consumption calorimeter Device to measure energy release rate in fire.

Oxyhemoglobin Oxygen-bearing hemoglobin.

Parts per million, ppm Concentration based on 10^6 parts of the mixture.

Performance codes Regulations providing for engineering analysis.

Piloted ignition Ignition of a flammable fuel-air mixture by a hot spot, spark, or small flame (pilot).

Pool fires Fires involving horizontal fuel surfaces, usually symmetrical.

Premixed flame A flame in which fuel and air are mixed first before combustion.

Products Chemical compounds produced by fire.

Pyrolysis The process of heating fuel to cause decomposition.

Radiation Heat transfer due to electromagnetic energy transfer such as light.

Radicals Short-lived unstable molecules such as OH.

Rankine (°R) Absolute Fahrenheit temperature scale, 460 + °F.

Reradiation The radiation reemitted from a heated surface.

Respiration minute volume, RMV Inhalation rate.

Shock wave Abrupt change in temperature and pressure due to a flow instability caused by speeds in excess of the speed of sound.

Smoke Gases, no longer chemically reacting, that emanate from the fire.

Smoke visibility, L_v Ability to perceive objects through smoke over a specific distance.

Smoldering A slow combustion process between oxygen and a solid fuel.

Soot Carbonaceous particles produced in flames.

Species Another name for chemical compounds, usually gases.

Specific heat Property that measures the ability of matter to store energy.

Spontaneous combustion A process by which combustion occurs after a self-incubation period between a fuel and oxidizer.

Stack effect Motion of air and smoke due to buoyancy, usually in a tall building.

Stoichiometric Refers to the amount of air needed to burn the fuel (and to combustion products formed).

Stoichiometric air to fuel mass ratio, *s* The ratio of air to fuel by mass needed to burn all the fuel to completion.

Surface tension A force within the surface of a liquid.

Synthesis Recombination of molecules.

Thermal conductivity The property of matter that represents the ability to transfer heat by conduction.

Thermal energy Energy directly related to the temperature of an object.

Thermal inertia A thermal property responsible for the rate of temperature rise, $k\rho c$.

Thermal runway An accelerating chemical reaction due to an imbalance between heat loss and energy production.

Thermocouple Device made of two dissimilar metal wires to measure temperature.

Thermodynamics The study of energy and states of matter.

Thermoplastic A synthetic polymer that usually softens and melts on heating.

Thermosetting Refers to a synthetic polymer that contains cross link bonds and usually chars on heating.

Turbulent Refers to randomly fluctuating fluid motion around a mean flow.

Underventilated Less than stoichiometric air is available.

Upper and lower flammability limits Concentration of fuel in air in which a premixed flame can propagate.

Vaporization temperature The temperature of a vaporizing fuel while burning, or needed to cause vaporization.

Ventilation factor The parameter controlling smoke flow rate through a door or window, $A_o\sqrt{H_o}$.

Ventilation-limited or ventilation-controlled State of a compartment fire where the air supply is limited; smoke gases will have nearly zero oxygen left; underventilated.

Viscous Refers to fluid friction.

Vitiation Refers to the reduction of oxygen concentration in air.

Volume fraction, *X* Species concentration based on volume.

Vortex A ring of eddies, or swirling motion.

Wavelength The distance traveled in one cycle, or the speed of light divided by frequency.

Wind-aided Refers to air flow in the same direction as the fire spread.

Work The movement of mass over a distance.

Yield The mass of product produced per unit mass of fuel supplied.

Zone model A type of computer fire model that approximates the fire conditions in a room as two uniform gas layers with a fire energy source.

Index